The Plant Hunter

The Plant Hunter

A SCIENTIST'S QUEST FOR NATURE'S NEXT MEDICINES

Cassandra Leah Quave

VIKING

VIKING
An imprint of Penguin Random House LLC
penguinrandomhouse.com

Illustrations on pp. 13, 115, 227, and 349 © Dr. Tharanga Samarakoon. Used with permission.

Species featured on the cover include American beautyberry (*Callicarpa americana*), St. John's wort (*Hypericum perforatum*), and Brazilian peppertree (*Schinus terebinthifolia*).

LIBRARY OF CONGRESS CATALOGING-IN-PUBLICATION DATA
Names: Quave, Cassandra Leah, author.
Title: The plant hunter : a scientist's quest for nature's next medicines / Cassandra Leah Quave.
Description: [New York City] : Viking, 2021. | Includes bibliographical references and index.
Identifiers: LCCN 2021011104 (print) | LCCN 2021011105 (ebook) | ISBN 9781984879110 (hardcover) | ISBN 9781984879127 (ebook)
Subjects: LCSH: Quave, Cassandra Leah. | Medical botanists—United States—Biography. | People with disabilities—United States—Biography. | Medicinal plants. | Ethnobotany.
Classification: LCC QK99.U6 Q38 2021 (print) | LCC QK99.U6 (ebook) | DDC 581.6/34—dc23
LC record available at https://lccn.loc.gov/2021011104
LC ebook record available at https://lccn.loc.gov/2021011105

Printed in the United States of America
1   3   5   7   9   10   8   6   4   2

Set in Baskerville MT Std
Designed by Cassandra Garruzzo

For the knowledge keepers and the wisdom seekers
who delight in nature's beauty and revel in its complexity

Contents

PART III

Medicine

Appendices

Author's Note

Throughout the book, I discuss my experiences in working with traditional healers and different systems of medicine. I also discuss specific uses of certain botanical ingredients. Plants have many names—even when being referred to in the same language. To avoid confusion over the common names of plants, I've also included their scientific names (genus, species, and family). An index of botanical nomenclature is included at the end of the book.

I have tried to re-create events, locales, and conversations from my memories of them. In order to maintain their anonymity, in some instances, I've changed the names of individuals and places, and in others, I may have altered some identifying characteristics and details such as physical properties, occupations, and places of residence.

The information presented in this book is not intended as a substitute for the medical advice of physicians. Readers should consult a physician in matters relating to their health and particularly with respect to any symptoms that may require diagnosis or medical attention.

The Plant Hunter

Prologue

In a wet grassy pasture bordering the swamplands of Florida, I stood with my team of six students, surveying what would be our work site for the morning. We'd all woken up early and driven the twenty miles into the countryside as dawn began to throw its rays over the swamp. This was the latest of many expeditions I had led for the students working out of my lab at Emory University in Atlanta, Georgia. Our mission: find plants.

The field was filled with invasive weeds—explosions of spiny soda apple (*Solanum viarum*, Solanaceae) with their hard, yellow fruits and prickly leaves—and clusters of the native feathery dogfennel (*Eupatorium capillifolium*, Asteraceae) that waved in the wind as if to greet us. Fifteen feet away, the green field turned to black muck, where a long stand of cypress trees, their knobby knees poking out from the water, stood as wizened sentries to the swamp. I stared at the scene a moment longer, wiping my brow. I hadn't even started to work and I was sweating like a pack mule in this April humidity.

As a professor and the herbarium curator at Emory, I'm dedicated to the discovery of new medicines derived from plants. My students and I were out in this swampland to collect some one hundred different plant species over the next two weeks; then we'd bring them all back to my lab to study. We'd spent the previous three months preparing for this trip, combing through historical records of Native American plant use, creating a customized field guide of the medicinal plants and their applications complete with identifying photos, and planning all of the logistical elements required of an intensive scientific expedition.

Finally, we were ready to go.

"Wait!" I said. "Let's change into tall boots."

Never underestimate Florida muck.

The students pulled on mud boots while I sat on our truck's tailgate. It would take me a little longer than them to get ready. Born with multiple congenital defects of my skeletal system, I've been immersed in medicine and science, in one form or another, right from the start. I've been subjected to scores of surgeries, beginning with an amputation just below my knee at the age of three, which have given me the ability to walk, using various models of prosthetic legs. Today, I wore my most stripped-down version—a metal pylon connecting the stump below my knee to a rubber foot shell. The prosthetic ankle didn't bend, making it almost impossible for me to put on a normal boot. Luckily, my handy husband, Marco, had installed a zipper down the back of this particular boot, so slipping it on was relatively simple. Once I was booted up, I checked the gear strapped to my army utility belt—a bowie knife, a set of pruning clippers called secateurs, a Japanese hori-hori digging knife, a handheld radio, and a .357 Smith & Wesson pistol.

We weren't there to hunt wildlife, but I knew better than to trespass on gator territory without a form of defense. My pistol was loaded with hollow points, the safety on. Few people are well acquainted with real Florida ecosystems—the inland swamps and wetlands where a picnic by the water can end with you becoming the meal—but this wild, dangerous, untouched part of the state was where I grew up.

Each student was equipped with handheld radios, clippers, and shovels. They spread out in pairs along the swamp's edge, entering from different angles like a big pack of wolves hunting down prey. I sent two of them off opposite our group toward the creek to wade through the shrubbery and tall grasses in search of the elusive blackberry bush, while the rest of us entered the swamp head-on.

Kim, an energetic young woman who'd been with the lab for a year now, was the first student to call me over. I used my hiking stick to help me slog through the rough terrain, my boots slurping and squelching in the muck. She pointed to a small herb, about a foot tall, with clusters of dainty white flowers along its distal stalk. "Hello, gorgeous," I said, touching it with my fingertips. I flipped through the pages of our field guide to confirm my hunch.

"Great find!" I said with reverence. "It's from the Saururaceae family—it only grows in this kind of swampy habitat." Its name is lizard's tail (*Saururus cernuus*) and its roots are used by the Cherokee and Choctaw people to make a poultice to apply to wounds; the Seminole use the whole plant to treat spider bites.

With our day's first find confirmed, the students around me knew the drill. I clipped a specimen to be included in our expanding herbarium collection, while Kim and the others began carefully collecting a bucketload of the plant so we would have enough for extraction back in the lab. We pressed the individual clippings into sheets of newspaper labeled with their name, collection number, date, and location. Layers of cardboard and blotter paper were then added and stacked inside of a wooden-slatted plant press, which could be squeezed tightly together for the drying process to yield flat samples for long-term storage in the Emory University Herbarium.

We continued this way—search, identify, collect—like some crazy combination of a platoon on a reconnaissance mission, off-the-grid campers collecting foodstuffs, and shoppers wandering the aisles of the world's biggest and wildest Walmart. After an hour or so of strenuous work and more than a few sought-after plants located, I heard Josh on the radio triumphantly announce those two precious words I had been longing for: "Found it." A jolt of joy shot through me. They had found the blackberry bush.

At the heart of science is the unalloyed thrill of discovery. Every plant I seek out during fieldwork is like buried or long-lost treasure I'm on the hunt for, but this type of blackberry bush, known as a *Rubus* species, was the mother lode for this journey. I was so eager to see it, I nearly slipped in the muck—clomping and stomping—as I made my way to Josh. With well over three hundred different species scattered across the earth, *Rubus* has been used for centuries by Native peoples to treat diarrhea, sexually transmitted diseases, and skin infections. I was hoping to add some North American species to my roster of studied plants, because I'd already examined one species native to Italy that demonstrated extremely promising antibacterial properties. It was a biofilm inhibitor, which means it blocks bacteria's ability to stick to surfaces, making the bacteria more vulnerable to attack from antibiotics or the immune system. If we collected even more *Rubus* species, we could learn more about their potential utility against one of the most troubling developments since the discovery of antibiotics: antibiotic-resistant bacteria. This is my enemy, my archnemesis, and what I do is develop weapons to fight what may become the next major war in humanity's struggle to survive.

MORE THAN SEVEN HUNDRED THOUSAND people across the globe currently die each year of an infection that is untreatable due to antimicrobial resistance (AMR). By the year 2050, the number of AMR deaths is projected to exceed 10 million—that's nineteen people dead per minute—overtaking the number of projected deaths due to cancer (8.2 million), diabetes (1.5 million), diarrheal disease (1.4 million), and road traffic accidents (1.2 million). To put this in perspective, most school buses can hold about fifty adults. We lose about thirty-eight school bus loads of people *every day* due to AMR infections. Thirty years from now, we expect to lose about twenty-two busloads of people *per hour* due to AMR infections. AMR is like a freight train racing at full speed toward the end of the line with no tracks . . . and no breaks. We have a serious problem.

Even scarier is that since the 1980s no new chemical classes of antibiotics have been discovered and successfully brought to market. The discovery and development of new chemical structures with the right type of antibacterial activity and human safety profiles is not a trivial process—it takes enormous investments of time, expertise, and funding. Consistent funding, to be exact. In the past decade, out of the fifteen new antibiotics that won hard-earned FDA approval, five have essentially been shelved as the companies that developed them either filed for bankruptcy or were sold. We're facing a double crisis in the battle against superbugs: the loss of effective antibiotics and the cataclysmic failure of the economic model that supports their discovery and development.

How did we get here?

The microscopic culprits that cause debilitating or deadly infections vary in form, lifestyle, and habitat. Some are viruses, others fungi or parasites. Many are bacteria. Bacteria are single-celled organisms that persist everywhere in the environment—they even live on our skin and inside our body. If all large life-forms disappeared—whether plant or animal—the world would be filled with their ghostly outlines, composed of microbes tracing their insides and outer surfaces.

While most of the time bacteria are harmless (*microbiome* is the term for the collection of the bacteria and other microbes that are naturally occurring in and on your body), others act as opportunists and transition from harmless interlopers to wily invaders of the body, causing minor and major infections. This is where antibiotics come into play. And it's why I waded through weeds and vines, kept my peripheral vision trained for any restless gators, searching for that blackberry bush. This *Rubus* species could act as a new form of antibiotic.

It's difficult for us now to imagine what life without antibiotics would be like. Yet, incredibly, antibiotics as we know them have been around for less than a century—a mere blip in time when one considers the history of humankind. In 1900 in the United States, during the pre-antibiotic era, the top three leading causes of death were infectious diseases: pneumonia, tuberculosis, and diarrhea or enteritis (inflammation of the small intestine). The discovery of penicillin in 1928 and its mass production fifteen years later is rightfully known as a game changer in medicine. It might even be *the* game changer. As the discovery of fire in prehistoric times changed the way humans prepared and consumed food, penicillin pushed us into a new age of medicine. The advances made by Alexander Fleming, Ernst Chain, and Howard Florey (1945 Nobel laureates in Physiology or Medicine), as well as Dorothy Crowfoot Hodgkin (1964 Nobel laureate in Chemistry), set us on a path of discovery unlike any before. Patients on the verge of death were brought back to the realm of the living.

Penicillin works by interfering with the way that bacteria grow and divide. Bacterial cell walls are made of peptidoglycan, and by blocking the linkage of peptidoglycan units during the growth process, penicillin causes the bacteria to rapidly fall apart, spilling their inner contents out into their environment much like a water balloon with a leak; this kills the bacteria. Following this innovation, others quickly followed, marking the beginning of the golden age of antibiotic discovery in which scientists scoured the earth for other microbes able to produce these wonder drugs.

Not long after their introduction, though, antibiotics became subject to misuse. Penicillin was made available to the general public in 1945, and people began to use it for minor complaints, including viral infections, for which it had no actual utility. Everyone wanted access to the newest miracle of medicine. It wasn't uncommon to take the drug just long enough to relieve symptoms and then stop. This leads to antibiotic-resistant bacteria. Even Alexander Fleming, the Scottish physician and microbiologist who first noticed the antimicrobial power of a simple speck of the penicillin mold, knew that resistance was bound to become a problem for this new life-saving drug. In a 1945 speech, Fleming cautioned the audience on the dangers of overusing this wonder drug: "The greatest possibility of evil in self-medication is the use of too-small doses, so that, instead of clearing up the infection, the microbes are educated to resist penicillin and a host of penicillin-fast organisms is bred out which can be passed on to other individuals and perhaps from there to others until they reach someone who gets a septicemia or pneumonia which penicillin cannot save." His prediction on this trend in acquired resistance and the spread of resistant bacteria from one person to another has unfortunately proven all too prescient.

The challenges to antibiotic research and development shouldn't be underestimated. This line of work is really hard. Almost all of the classes of common antibiotics in clinical use were discovered by the early 1950s, before we even knew the structure of DNA. Despite the enormous advances in molecular biology, biochemistry, pharmacology, and many other areas, we have been struggling for decades, mostly unsuccessfully, to find novel antibiotic compounds. As Dr. Tom Dougherty, an expert who spent his career working in pharma on antibiotic research and development, once said to me: "If you don't have a high tolerance for failure, maybe you should quit this field and do something less hard, like rocket science. At least they have largely succeeded since the 1950s."

Further complicating matters, big pharma is getting out of the antibiotic game, as it has not been reaping ample enough returns, leaving a massive chasm in its wake.

We need new therapies, and we need them now. That's where I come in.

Of the estimated 374,000 species of plants on earth, records exist for the medicinal use of at least 33,443. While some of these plant-based medicines have been preserved in books like the Roman physician Pedanius Dioscorides's first-century *De materia medica* and the nun Hildegard of Bingen's twelfth-century *Physica*, many other natural medicines exist in oral traditions, practiced by lines of shamans or medicine men (and women). *Thirty-three thousand species!* That means that around 9 percent of all plants on earth have been—and in many cases, continue to be—used as a major form of medicine for people. There's only one thing more amazing than the fact that humans, over thousands of years of trial and error, across cultures and landscapes, have sorted through hundreds of thousands of plant species to determine which ones might soothe headaches and upset stomachs, ward off hunger or thirst, and relieve pain, and have come up with at least 33,000: that we haven't *really* bothered to study the vast majority of these plants.

Shockingly few of these species have been rigorously investigated under the lens of modern science for their full pharmacological potential—no more than a few hundred, if that. That means that out of the 9 percent of plants of which documented pharmacopoeia exists, less than 5 percent of these have ever even entered a lab. While traditional healers know these medicines work, we have little scientific understanding of *how* they work. We've only scratched the surface! And yet of the plants that we have studied and refined, the results have been utterly amazing.

Plants serve as the derivations for an array of life-saving and health-improving medicines we all now take for granted. Ever taken an aspirin? Thank a willow tree (*Salix* species) for that. Ever received a shot of numbing medication at the dentist's office prior to a procedure? That was originally discovered in a plant, too—the coca plant (*Erythroxylum coca*, Erythoxylaceae) from the Andes. What about a painkiller for surgery? Morphine from opium poppy (*Papaver somniferum*, Papaveraceae) really takes the edge off the post-op pain. Many of the groundbreaking drugs for cancer came from plants—Taxol from the Pacific yew tree (*Taxus brevifolia*, Taxaceae) in the northwest United States; etoposide from the mayapple (*Podophyllum peltatum*, Berberidaceae) in the eastern United States; vincristine from the Madagascar periwinkle (*Catharanthus roseus*, Apocynaceae); and camptothecin from the Chinese happy tree (*Camptotheca acuminata*, Cornaceae). If you've ever traveled to a malaria zone or had the misfortune of contracting the disease, the first treatment was found in the Amazon, quinine from fever tree bark (*Cinchona* spp., Rubiaceae); and later, from China, artemisinin from sweet wormwood (*Artemisia annua*, Asteraceae). A Nobel Prize was even awarded for that discovery.

Until the dawn of synthetic pharmaceuticals in the late 1800s, all medicines were derived from plants, animals, and minerals, whether mixed, compounded, extracted, cooked, or raw. We have lived in an age of synthetic drugs for less than 150 years, but that has been long enough for us to forget where they came from.

Today, most of us are far removed from the process of growing or wildcrafting foods and medicine. Instead, we rely on goods sourced from industrial agriculture and pharmaceutical factories. This has been a great boon for humanity in many respects—more food for more people, cheaper drugs more readily available—but this great progress has come at a cost. Our loss of familiarity with the powers of the natural world, combined with decreasing access to land on which to grow or collect these resources, has fostered a widespread and dangerous decline in our understanding of the critical role that it plays in our health. Having lost our connection to the natural world, we often see ourselves as separate from—or even opposed to—the forces of nature. But humans are very much a part of nature, and always have been.

Unfortunately, under current reductionist scientific paradigms, many scientists focus exclusively either on man-made compounds or on the "quick fix" of simply making structural modifications to existing drugs. While such modifications can be useful for a time, they leave untapped a vast repository of potential medicines that various traditional cultures have shown to have great promise in treating ailments. What's more, the approach to drug discovery most common in the West prioritizes a single compound or a single biological target, which means that studying the complexity of network pharmacology inherent in plants is often dismissed out of hand—deemed too complicated, too time consuming, and too expensive.

But there is another way. Ethnobotany is the scientific study of how humankind interacts with the environment in the procurement and transformation of plant materials into food, building materials, tools, and for me, especially, medicine. The term *ethnobotany* was first coined in 1895 by American botanist John Harshberger. Ethnobotany emerged as a formal scientific discipline in the twentieth century, though botanists, physicians, and scholars have long been captivated by documenting the properties of useful plants. The rich history evidenced in ancient herbal texts and scrolls written in Greek, Chinese, and even Egyptian hieroglyphs is testament to the human fascination with recording and saving knowledge of useful plants for sharing with others.

Plants affect us in immeasurable ways. They are what we use to nourish and clothe our bodies (cotton T-shirts), build our homes (durable timber), and enrich our lives with art (plant dyes) and music (cane plants used for reeds in wind instruments). In short, ethnobotany is the science of survival. As an ethnobotanist, I know all too well that looking through nature to discover new medicine is difficult, time consuming, and expensive. However, not only do we benefit by doing so, as nature can and has yielded amazing results, but we've never needed it more. I believe we will face a grave, deadly, and costly crisis if we don't act now to increase our study of plants to develop new ways to treat infections. Indeed, our very survival may depend on it.

I USE AN ETHNOBOTANICAL APPROACH to identifying drugs from nature to discover new ways to treat antibiotic-resistant infections. That's what my team's sweaty trek through the Florida swampland was all about. My work has taken me to the flooded forests of the remote Amazon, rattlesnake-infested pinelands of Georgia, isolated mountaintops in Albania and Kosovo, the rolling hills of central Italy, and rocky isles arising out of the Mediterranean. I now have more than 650 medicinal species and 2,000 extracts in my chemical library to use in my research group's drug discovery work. I've traveled by dugout canoe where rivers are the highways of the forest, quad bike over desert dunes, four-wheel-drive up rugged mountain roads, mule where it was too steep for me to climb, airboat through the Everglades, and ever so slowly on foot to hunt for medicinal plants and interview the people who use them. I've done this all in the hope that we as the human species might work with nature to help ensure our own survival.

This quest is deeply personal for me. I was born with a constellation of orthopedic defects including the absence of several bones in my right leg and foot, and at the age of three I nearly died from a staph infection. Staph infections kill more than twenty thousand people in the United

States annually. Luckily, an antibiotic was used, and the infection was defeated.

Modern medicine saved my life, as it has done for hundreds of millions of people over the past century. My quest is to expand the efficacy and impact of tomorrow's medicine by broadening our knowledge base and research pipeline. I aim for inclusivity in my approach, considering the insights of multiple divisions of science, the arts, and humanities. Humbly, I listen and learn from the elders who offer the wisdom of generations of knowledge gained and passed down over the centuries, unique to their cultures and ways of knowing health and practicing medicine. With vigilance and persistence, I believe we can learn a great deal from people, places, and experiences that have been overlooked or discounted. The bacterium that nearly ended my life is the same one that the blackberry extract I've studied in my lab has proven effective in fighting.

This is the story of my journey through the struggles of disability and infection, to my fascination with science and dedication to it, to my development as a scientist and explorer, through marriage, motherhood, and finding my own place in a male-dominated field of science to seek out new ways to win the war against infectious disease. Through the many surgical interventions that I underwent, I gained decades of personal experience with the strengths and failings of modern Western medicine. It is through the lens of these experiences that I came to understand how humans are deeply tied to nature, how nature can provide clues for the next generation of advanced medicines, and how we may overcome antimicrobial resistance by targeting the ability of bacteria to cause harm, rather than just trying to kill them outright. I believe nature can save us.

PART I

Nature

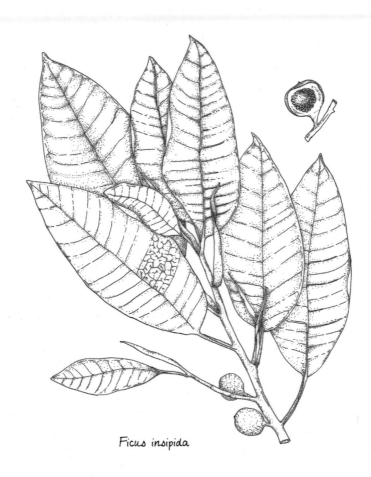

Ficus insipida

CHAPTER 1

My Leg and the Wilderness

You only live twice:
Once when you are born
And once when you look death in the face.

IAN FLEMING, *YOU ONLY LIVE TWICE*, 1964

I'm descended from a long line of Quaves, running all the way back to Juan de Cuevas, born in Algámitas, Spain, in 1762, who settled in what is now known as Harrison County, Mississippi. He married into the French Canadian Ladner family that had settled in the region, and they made their home on Cat Island, raising twelve children. To this day, their descendants carry a number of similar surnames—Cuevas, Coueves, Quave, and Queve. The name on my line of the family is pronounced *kwave* (rhymes with *wave*). We've always been people of the land.

My father, Raymond, grew up on Quave Street in Biloxi, which, at one time, was entirely inhabited by close relatives. Daddy's father, J. L., was a stumper. Back in the 1920s, colossal longleaf pines (*Pinus palustris*, Pinaceae) throughout the south were clear-cut and used to build the homes and businesses of small towns that popped up along the coastline and interior of Florida. These evergreen trees sport the longest needle-like leaves of any pine in the world and proudly stand between eighty and one hundred feet tall. After trees were felled, the stumps stayed behind. J. L. and his sons used bulldozers to push the stumps out of the ground and then blew them up with dynamite, blasting them into small enough pieces for grinding at the stump mill. The Hercules Powder Company had such a mill located on the Peace River in DeSoto County, Florida. After being washed, the stumps were ground into chips and steamed to extract turpentine and other by-products like nitroglycerin and black powder.

Stumping was an essential part of clearing the land following the timber harvest and creation of arable land for agricultural use. It was not easy work. My uncle Tommy lost several fingers playing with a hammer and a dynamite blasting cap as a kid, and another close member of J. L.'s work crew, Bo, died when a chain carrying heavy equipment broke and crushed him in the semitruck cab. Daddy grew up working outdoors, welding together pieces of scrap metal both as an artistic outlet and as a means to repair pieces of the business's bulldozers, tractors, and excavators. After the family moved from Mississippi to Florida, Daddy and his brothers became well known for their wild days of drag racing cars and trying to outrun the police. One of my uncles even served on a chain gang after being caught by a policeman.

At the age of twenty in 1969, Daddy made his first trip overseas, trading the swamplands of Florida for those of Vietnam. He served in the First Infantry, Eleventh Brigade, Company B, Third Battalion, Americal Division, Central Highlands. He trekked through miles of jungle recently defoliated by dioxin, or Agent Orange—a powerful herbicide sprayed by US military forces to eliminate forest cover and crops of the North Vietnamese and Vietcong troops. In late July 1970, his machine gun locked and loaded, he was on top of an armored personnel carrier (APC) with four other friends in his squad, their eyes on alert as they surveyed the horizon, sweat dripping down their backs. Six other members of his platoon rode inside the APC. They were positioned in North Vietnam, north of the Quảng Ngãi Province, with orders to move to another region where the Vietcong had been spotted. It happened fast. The startling sounds hit before the explosion could even register in Daddy's mind. The road they were traveling on had been rigged with booby traps—land mines hidden beneath the deceivingly plain soil.

Only he and the other four who rode on top of the APC survived. A chopper arrived quickly enough to medevac them to a military evacuation hospital in My Khe, a beach city in South Vietnam known as China Beach. Daddy didn't lose any limbs, but he suffered from back pain for the rest of his life where his vertebrae had been impacted. He received a Purple Heart for his service and was shipped back home. It was there in Arcadia, Florida, that he met my mother, Cissy. At the time Raymond had been in Vietnam, Cissy was attending Samford University in Birmingham, Alabama, first training in the field of biology and medical technology. While spending a summer interning at G. Pierce Woods, a medical facility for mentally disabled and psychologically ill persons in Arcadia, she and Daddy started dating. Tall and beautiful, with a youthful face and long straight chestnut hair that reached her waist, she was quite a catch.

My parents decided to start a family in that small town of Arcadia, located in southwest Florida in DeSoto County—about an hour's drive inland from the better-known coastal towns of Sarasota and Fort Myers. Daddy, like his father before him, worked in the stumping business, later expanding his work to larger-scale land-clearing operations in support of the booming agricultural industry (beef and dairy cattle, orange groves, and farms for tomatoes, strawberries, and watermelons). Momma worked in the school system as a special education teacher. They were happy and healthy and in love.

And then I came along.

. . .

I WAS BORN WITH THE LOWER PORTION of my right leg underdeveloped, the foot bent in a strange angle at the ankle, the tibia above it short, and the fibula totally missing. Anklebones were missing, too. Imagine two legs, but one substantially shorter than the other, with the bottom of thzot only reaching the calf muscle of the longer leg. The doctors were baffled: there had been no indication of my birth defects during my mother's pregnancy. But this was 1978, a time before ultrasounds were a normal part of pregnancy checkups. My first few months of life were a constant series of visits to various doctors, putting a strain on not only my parents' emotional well-being but also their finances.

As I grew, doctors explained to my mother that the pseudarthrosis (false joint) in my ankle combined with the growing limb-length disparity would make walking a serious challenge. Even if I wore braces to give my leg support, my pseudarthrosis would put me at risk of frequent fractures in my foot. To make matters worse, one medical resident predicted that in addition to my physical disabilities, there was a very high likelihood I'd also suffer from severe mental impairment, common in children with constellations of developmental defects—this was the one prediction that didn't materialize even as the others did. In short, they learned that a lifetime of risk, relative immobility, and imminent pain was what awaited me. My parents were crushed.

By the time I turned three, they needed to make a choice. After reviewing the stacks of medical records and considering the doctors' recommendations, my parents decided the best path forward was below-the-knee amputation: this would make walking easier for me with the use of a prosthetic leg. It wasn't an easy decision; even as they weighed the pros and cons of the choice, family members accused them of planning to mutilate their child. Tensions ran high.

By August of 1981, just two months after my third birthday, a surgical team successfully removed part of my leg. I was left with a nice residual limb below the knee complete with a thick pad of fatty tissue beneath the bone to fit into the socket of a prosthetic leg. After almost two weeks in the hospital, I was ready to go home. Everything went perfectly.

Three days later, back at home, Momma smelled something rancid. It wasn't coming from the compost or the trash can; it was coming from me, from my stump. The discharge nurse had said not to remove the bandages under any circumstances, but something deep inside of Momma told her there was a serious problem. The smell reminded her of infected wounds of sick horses from her youth, and so she began the slow process of unraveling the long spool of outer Ace bandages, as if she were unraveling a mummy from its protective covering. Very gently, she peeled off the next layer of thick cotton padding, when liquid started to drip out, thick and yellow like a snotty vanilla pudding. After lifting a stack of white nonstick bandages, the last barrier to the stump, she gasped as she watched a puddle of thick fluid, bits of fat and rotted tissue from the bottom of my calf muscle, fall to the floor. A long piece of white bone jutted out from the mess. Her instincts had been correct: my limb was rotting due to an infection. She rushed me back to the hospital.

$N_{ooooooo}$!" I SCREAMED, naked and squirming in the grasp of Momma's arms as a nurse assisted. "Don't put me in the blood!" I sobbed, plump tears rolling down my reddened, puffy face.

We were in the whirlpool facility at Sarasota Memorial Hospital, and this was the first of many sessions of whirlpool therapy that I would have to undergo to treat my infected leg. It followed the debridement of my stump—a process by which the doctor had scraped the rotting flesh away. The drab room featured a floor of gray tiles punctuated by a few circular tubs on raised platforms. The smell reminded me of the operating room I'd been wheeled into weeks earlier, caustic and heavy in the air, making me cough between my sobs. Each tub was filled with Betadine solution—blood red in color—and bubbling from the jets that spun the solution around. To three-year-old me, they were putting me in a literal bloodbath.

Eventually, Momma won the battle of wills, as she often did, and I sat in the tub sobbing during the treatment. It wasn't painful—I was just convinced it was a vat of blood. An old lady who was also in the whirlpool center for therapy brought me a big yellow ducky the next day, and that eased the way for my future treatments over the next few weeks. Now it was just me and ducky in the bloodbath together.

Had it not been for Momma's foresight—indeed, if she had heeded the nurse's warning and not unwrapped the bandage when she did—I would have died. The infection had already ravaged my skin and soft tissues. I not only had gangrene but also a staph infection that had begun to enter the bone, forming a recalcitrant biofilm in the tissue—a disease known as osteomyelitis. I was hours away from the infection spreading to my bloodstream and then leading to sepsis and multi-organ failure.

Knowing what I do now about staph infections, I thank my lucky stars that it was a 1980s strain of *Staphylococcus aureus*, and not yet a multidrug-resistant HA-MRSA (healthcare-associated methicillin-resistant *Staphylococcus aureus*) prevalent in the nineties—and certainly not one of the horrifyingly aggressive CA-MRSA (community-associated MRSA) isolates that came soon after. Staph is devilishly cunning not only at spreading throughout the body, but also at forming a plethora of different types of infections in the bone (osteomyelitis), heart (endocarditis), bloodstream (bacteremia and sepsis), brain (abscess), and skin (soft tissue infections, even necrotizing fasciitis, or "zombie skin").

After three weeks in the hospital, I was sent home, again wrapped in padded stacks of bandages, and my parents watched closely for any other signs of infection. By mid-November, I was back in the hospital for another surgery. They needed to clip the bone back and stitch what was left of my flesh around it to create a stump that I could eventually walk on. As with the prior surgery and every surgery that was to come thereafter—there was fear, risk of death, and always the pain. But for now, I was in the clear. At the age of three, I had looked death in the face. There was more for me to do with this life. I was not ready to go yet. *You only live twice.*

EVEN WITH MY DISABILITY, I loved playing outdoors, whether exploring dense oak hammocks or the piney woods and palm thickets near our house with my dog, Spot, or playing with my younger sister, Beth, on a massive mountain of dirt Daddy had built in the backyard. But it was looking through a microscope that transported me from the macroscopic world of watching insects march around the forest floor, or observing deer sniff out a grassy meal in the field, to somewhere altogether new and magical. To see something move under that microscope was to enter another dimension, a world I never knew existed, one just waiting for me to explore. This was when I truly fell in love with science and found an outlet for all the pent-up energy that seemed thwarted by my disability.

In third grade in 1987, I participated in my first science fair. My question was simple: What lives in a drop of pond water? Momma borrowed a microscope from the high school for me to use at home for the project. Collecting the samples had been challenging on crutches, but Spot and I had made it out to the front pond and back with a plastic cup full of pond water in hand. My goal was to look at multiple samples of water droplets, draw and identify the creatures I saw, and count how many of each could be found. Simple, elegant, and very exciting to me.

Underneath that microscope, I discovered that pond water teems with life. Translucent bodies twisted and swirled beneath the bright light. I bent over the microscope for an even closer look, knocking a pair of crutches that had been leaning on the kitchen table onto the tiled floor. I couldn't be bothered to pick them up. This was too important. I twisted the knob to click a higher-resolution lens into place, fine-tuning my view with the knob in my other hand.

"It's covered in hair!" I shouted.

No one was there to hear me.

The paramecium swam around inside the drop of pond water, its elongated oval body surrounded by short little hairs: they wiggled to propel it into motion. I picked up my colored pencils to draw it in my notebook before turning my attention to another creature floating about under the microscope's lens. This one was not in any single particular shape, but ambiguous, shifting to form armlike projections as it slid from one side of the viewfinder to the other. I flipped the pages in my encyclopedia.

"Aha!" I said. "You must be an amoeba!"

I jotted this down in my notebook and patiently drew its globular-like shape on the page. Daddy came in through the back door to the kitchen, knocking the dirt off his boots on the outside steps, returning from work on his heavy equipment out at the barn.

"What are you doing?" he asked as he passed by to the kitchen sink to wash up.

My head popped up from the microscope.

"Did you know that there are paramecia *and* amoebas in the front pond?"

"That's cool," he replied with interest.

"I can't wait to see what's living in the ditch water. I'm hoping to see a spirochete!"

Spirochete! Amoebas! Protozoa! Algae!

It was all love at first sight (at ten times the power of the human eye).

After the data from my pond-water experiment was recorded, I went to work organizing my display board—hypothesis, methods, materials, results, conclusion, and bibliography. I spent long afternoons sprawled out on the living room floor, Spot panting nearby, as I carefully assembled all my project's components. Weeks later, I unfurled it in the school library for review by the judges. Nervously, I walked away on my crutches and looked at what other kids had done—foaming model volcanoes, growth studies on bean plants, and the like. When I returned to my display, there was a little blue ribbon pinned to my board. First place in my category! I had won. I'd never won anything before.

THROUGHOUT ELEMENTARY SCHOOL, I had to have surgery once or twice a year to revise my stump: I had the unlucky habit of growing bony knifelike protrusions called bone spurs on the tip of my tibia as I hit growth spurts. Following each surgery, I'd be stuck indoors until the wound healed enough to withstand my wet and dirt-covered play activities. Those long, cooped-up days passed by in slow-moving minutes of misery. Time dragged on—boredom became frustration, frustration became anger, and anger became envy of the able-bodied kids, like my sister, Beth, and the other kids in our neighborhood. I gazed out the window, longing to be outside playing with them. Although I didn't consider it at the time, those years must have been tough on Beth, too. Four years my junior, she'd known me only postamputation. So much attention was focused on me and my ongoing health challenges that she was left on the fringes, looking in just as I was gazing out.

School absences piled up from doctor's appointments that were hours away. I hated hospitals, my second home, and I missed out on a regular routine. It was hard to convey my everyday experience to my able-bodied peers whose experiences were nothing like mine.

"What did you get for Christmas?" kids at school would ask.

"Cherry Popsicles and a morphine drip!" would have been the truthful response.

"Did you go anywhere for summer break?"

"Yes! Back and forth to the hospital!"

Momma sometimes let me cry at night, holding me as I asked her through tears why I didn't have a normal leg. She was firm in her belief that I was like any other girl. She never drew attention to my leg and never let me off the hook. "You can do anything you want," she would say to me.

While Momma coached me to push past the many boundaries I encountered, Daddy taught me how to transform my ideas into something tangible. I've always been a dreamer, but it was Daddy who showed me how to become a doer. Whether it was a design I'd dreamed up for a treehouse or an invention of some practical crutch-holders to clamp to the table and keep my crutches from falling to the floor, he took my designs to the barn, where I watched him saw, hammer, or weld these ideas into a real product. Watching him work and tinkering with tools beside him, I saw how true Momma's words were. I *could* do anything I wanted with a bit of creativity and hard work.

I listened to Momma, her faith in the face of my hopelessness, and I resolved to live up to her expectations. I wanted to do science. That would be my sport. Encouraged by my success at the microscope in third grade, and driven by a curiosity to understand more about the invisible world, I decided to pour all of my energy into it. Science fairs became my softball or soccer tournaments, winner take all.

Following my pond-water experiment, I decided to examine spit. My question was, Who has the dirtiest mouth? Human, dog, horse, or cow?

As any good scientist would, I started with my own saliva. Spot and my horse, Sequoia, donated saliva, too. The cow, however, had other plans. I spent hours chasing her around on one leg with crutches, test tube bouncing about in the front pocket of my shirt. Spot tried to help as well, but he and the cow were not on the best of terms and his incessant barking didn't help. This was my first hard lesson in science: things rarely go as smoothly as planned. I removed the cow from the project, shrinking my sample size. Of the three of us, I was surprised to find that it was my saliva sample—not Spot's or Sequoia's—where the most microbes lived.

What I was able to see that day under the microscope lens was just the barest glimpse of the complexity of life carried in our mouths and those of our favorite pets. Much more advanced tools—which would not come to science until decades later—would reveal the incredibly important role that our microbiome plays in maintaining a healthy body. Advances in genetics, sequencing technologies, and computing capacities have opened up whole new avenues for investigating the ecosystems of the human body, with all of the myriad microscopic members of the site-specific community. For me, I was beginning to appreciate more and more the vastness of life and the diversity of creatures hidden from the naked eye but ever present around us and especially in our bodies.

As I went from elementary school to middle school, my science fair experiments advanced from simple observations of life teeming in pond water and saliva from my home kitchen to more experiments undertaken on human pathogens in the clinical microbiology laboratory at the hospital in the neighboring town of Punta Gorda. One particular pathogen caught my attention—a deadly strain of *E. coli* (short for *Escherichia coli*)—and I wasn't the only one. While this particularly deadly strain of *E. coli* had been discovered back in 1982, it made national news in the late 1980s and early 1990s, leading to widespread hysteria over contaminated beef. In one of the biggest outbreaks, in 1993, undercooked hamburgers at a fast-food chain made more than seven hundred people across multiple states sick, even killing four children. Americans had long thought their processed food was safe—and, suddenly, it was anything but. Special TV reports, banner headlines, magazine articles, new dietary habits—this was a big, scary story, and I felt drawn to it.

I kept turning it over in my mind: a family of four goes to a drive-through hamburger joint—like millions of Americans do every year—and orders burgers for everyone. Mom, Dad, and ten-year-old Bobbie have stomachaches later, maybe some diarrhea. Four-year-old Tim, however, is sleepy and cranky, with a severe stomachache. This is followed by bloody diarrhea. Things get worse for Tim. The parents bring him to the hospital, where he develops kidney failure and dies within days. This is the stuff of nightmares for any parent, and it was happening because a supply of beef had been contaminated with a new, virulent strain of the gut bacterium *E. coli* O157:H7. Simply put, eating an undercooked hamburger tainted with this pathogen could become a death sentence, especially for the very young or the elderly.

My hometown of Arcadia has a long history in the beef cattle industry—complete with rodeos and auctions—and this contaminated beef problem was a fascinating riddle for me to investigate. Plus, it was *significant*, bigger than spit or pond water, since Arcadian cattle ranchers' livelihoods were being jeopardized by news stories about infected meat. This was a real-world problem that I could sink my teeth into (not literally, of course).

Wanting to better understand the link between ground beef and the deadly *E. coli*, I built my next middle-school science fair project around this topic. My question was to the point: Which vendors produced the cleanest meat?

The deadly *E. coli* bacterium is introduced to beef during processing; it comes from contact with cow feces present on the animal hide when the meat is butchered and ground up. I went to each of the grocery stores in town, representing major brand chains, and then also to a local butcher, purchasing samples of ground beef. Outfitted in my standard personal protective equipment (PPE)—lab coat, gloves, and goggles, which were now second nature to me—I first used a precision scale to weigh out a gram of ground beef taken from each of the different stores, placed them into a test tube with sterile saline (lightly salted phosphate-buffered water), and transferred some of the dilution of the meaty liquid to a petri dish containing nutritional blood agar that *E. coli* loves to grow on. After I gently spread the liquids across glass plates, I placed them in the incubator, where the bacteria would grow.

The next day, I eagerly opened the warm, humidified incubator only to be met with a strong whiff of putridity—these are fecal bacteria, after all. Using a handheld tally counter, I began the long and arduous task of visually counting the number of creamy, glistening colonies of bacteria on each plate and recording the numbers in my notebook. As I worked through each stack, a pattern had begun to emerge, and I became more and more excited. The results were clear: the national chain stores had a much heavier bacterial load than the beef prepared by the local butcher.

Every subsequent year, I expanded my exploration of this pathogen, first looking for ways to kill it. Working alongside a hospital radiologist who had agreed to assist me, I took my samples of beef and placed them in the all-too-familiar room of an X-ray facility. As I carefully lined up each sample under the red X of the laser beam on the metal bed, my mind drifted back and forth to the many times my body had lain prostrate on a similar hard bed, with a technician lining up the X over my hip, leg, or whichever part of my deformed bones needed to be assessed at the time.

Once satisfied with the arrangement of the sample, I scurried behind the protective lead wall barrier of the room with the radiology technician and we would zap the raw meat with the planned dose. I cautiously compared my control groups (untreated with radiation) to those under radiation to find the number of glistening colonies of bacteria growing on the deep red blood agar plates. Once again, the results were clear. The radiation-treated meat had significantly lower bacterial loads than my controls. I was intrigued by the possibilities that irradiation of food could have for ensuring the delivery of safe (pathogen-free) products to customers.

In high school, my projects shifted from foodborne bacteria to those isolated from patient samples. I began looking into the antibiotic-resistant profiles of different clinical isolates, collecting *E. coli* from the urinary catheters of infected hospital patients. Implanted devices—particularly any that remained in the body for long periods—are especially prone to being colonized by bacteria and causing an infection. As expected, they were often multidrug resistant. The process of how resistance was acquired and the rates at which it could achieve this became a point of focus for me. To test it, I streaked the bacteria into dense lawns of growth on blood agar–filled petri dishes. When I exposed them to low doses of antibiotics and after twenty-four hours of growth in the incubator, I looked for mutants—single bacterial colonies that appeared as creamy circles in the "kill zone" where no other bacteria survived. The appearance of resistant mutants was a one-in-a-million kind of chance, but here is the thing: the generation time (or doubling time) of *E. coli* under optimal growth conditions in a laboratory setting can be as quick as fifteen to twenty minutes! This means that a one-in-a-million chance mutation under the selective pressure of the antibiotic in a population of millions of cells is not that rare after all.

THE LINKS I WAS DRAWING between microbes and the health of the humans they infected became a point of intense fascination for me. Yes, the microbes *were* interesting, but what really drew my attention, I realized, was the human cost of infection. It was, after all, part of my own story. I'd been a part of the patient-doctor relationship for years, but I'd never gotten a glimpse of what it meant to be on the other side, the care provider rather than the recipient. It wasn't fun to be a patient, but it seemed interesting to treat one—and to understand what doctors were up against.

I went through doctors like most kids go through teachers, and what I became aware of as my sample size grew was this: there was a world of difference between doctors who worked like car mechanics—find the problem, tighten the screw, order the part, pass along the bill—and those who treated me and my body as a living, breathing entity. In short, the difference between being treated as a problem and as a person. Most of my doctors treated me as an object, something to be poked, prodded, discussed, spoken over—and sometimes even an oddity to be shown to medical students and residents, like a bearded woman at a traveling circus. But Dr. Price was different. He was an orthopedic specialist whom Momma found to treat the growing length disparities in my femurs, the hip dysplasia I developed, and the scoliosis that soon followed. Dr. Price spoke *to* me, not *over* me. By the time I reached third grade, my right knee was about three inches higher than my left—so it was little wonder that my gait was so awkward, or that years later in high school, I would suffer from muscle spasms that felt like midsize bolts of lightning striking my back.

Dressed in his white medical overcoat, Dr. Price explained to me that he was going to adapt a procedure, usually performed for patients with dwarfism, to lengthen my limb. In the operating room, he made an incision into the upper thigh down to the bone, sawed the femur in half, and then, on the outer side of my right thigh, drilled two pins into the bone above and below the location of the break, with a single bar-shaped external fixator that was attached to the pins, running along my right thigh. Basically, by cutting the bone and very slowly separating the two ends, the bone-healing process would naturally, over time, fill in the gap, making my right leg catch up with my left.

All this fascinated me. I understood that a child's bones develop in growth spurts and that's how she becomes a grown-up, but the idea that my bone could be separated and would then fill in the gap seemed like magic, like the autotrophic algae producing their own food. Because my body was so often talked about around me as if it were a piece of broken machinery—fix this, do that—I became interested in knowing all I could about my own hardware. I went to the set of encyclopedias we had at home to learn about organs, muscles, and bones. When I couldn't find the answer written down, there was always someone I could ask at my next appointment.

Dr. Price loved to engage me about the specifics of his work, and he even explained that I was the first amputee he used this new technique on. His other patients to benefit from the surgery were achondroplastic (little person) children. Months later, when the procedure had completed its work, he let me do the honors of unscrewing the device from the bone. Momma took a picture of me just after, beaming with delight, the metal screws in my hands. With that one gesture, he had made me part of the healing process, not just the subject of a medical intervention as had always been the case before.

It was Dr. Price's way of practicing medicine, engaging with the patient as much as with his profession's knowledge, that led me to volunteer for the candy striper program at the single county hospital in Arcadia, a couple of blocks from my middle school. Starting in seventh grade after school, I worked on the main hospital floors to check on patients, bringing them fresh ice water, extra blankets, or pillows when needed. I ran samples from blood draws and urine to the in-house clinical analytical lab for testing. For my few hours of work, I got a free meal in the hospital cafeteria (they had surprisingly good brownies). The program rotated us through different sections of the hospital—main admissions, obstetrics, and, for select volunteers of the right age, the emergency room (ER).

By the ninth grade, I finally got my chance to see the action in a real ER. From the first day I entered those bustling, important hallways, seeing real medical care in action, I was hooked. Right then and there, I wanted to become a physician. As the yearbook editor, I'd often hit up Friday-night football home games, standing on the sidelines downfield from the cheerleaders, snapping photos of touchdowns and interceptions, before I left to go to the hospital ER. While I was putting my scrubs on, my classmates were putting on makeup or pregaming, partying in cow pastures with booze and bonfires, or drag car racing on back roads. Friday and Saturday nights were when the most trauma cases poured in—barroom knife fights, heroin overdoses, drunken car crashes—so that's when I wanted to be there. It was like I was living a double life: schoolgirl and science nerd by day, secret intern immersed in the adrenaline rush of the ER by night. I was the only teenager to volunteer in the ER; in fact, I was the only volunteer there those years—period. As strange as it sounds, I spent so much time there—an average of fourteen hours a week—that it truly became a second home to me, this time one I welcomed; and the medical staff, a second family that shepherded me through my teenage years.

I observed the full spectrum of life and death, how the body changes when it goes stiff with rigor mortis, the way the color drains from a body as blood stops circulating, leaving it an eerie blue. I did my best to provide comfort to the patients and their families. I saw people at their most painful moments—drunkards carted in with gunshot wounds, mothers-to-be mourning the loss of a pregnancy. But I also saw spouses rejoice after successful procedures and parents summon courage when their child was scared and trembling.

When I was sixteen, I looked on helplessly as a boy from my school entered the ER on a gurney. He had drag raced down a country road with his friends, ending with his car wrapped around a tree. The ER team strapped electric paddles to his chest as they fought to zap a current back into his heart in between the rounds of CPR. The boy's face remained mottled and blue—most of his bones were broken. He died. Seeing someone my own age pass away from something so senseless shook me. I couldn't breathe. I walked outside for some fresh air. The night sky was cloudy, the damp air making me feel as if the whole world were collapsing around me. One of the veteran paramedics came outside, too.

We stood in silence for a moment, both teary-eyed. He turned to me and said, "It never gets easier."

Watching the physicians at work, I learned how to suture a laceration and look for signs of infection in the following days. I observed how to put in an IV, draw blood, examine X-rays for signs of pneumonia, and insert a urinary catheter. I learned how to install a nasal tube to deliver activated charcoal and pump the stomachs of overdose or poisoning victims, extract impacted ear wax with an IV bag and tube for irrigation, and remove the weird things kids stick in their noses or adults stick in their rectums. Sometimes, the cases from the ER would go on to surgery, and the local surgeon, with the patient's permission, would allow me to watch. The operating room became a comfortable place for me. I'd spent a lot of time there already, but now I was standing up, not lying down. I resolved that one day I would become a surgeon.

WHEN TIME CAME TO APPLY to college, I was most interested in finding a school with a strong premed program. I had a full scholarship to any of the Florida state schools, but I wanted to light out for a fresh start, somewhere kids from my hometown weren't going. I was through with keg parties on the riverbanks and high school cliques. Ultimately the compact size of Emory University's campus and its proximity to the Centers for Disease Control and Prevention (CDC) won me over. I wanted to be close to the action.

My first year of college was a struggle. Unlike many of my new peers, I'd never had the opportunity to take an Advanced Placement or International Baccalaureate class, and while I was used to being a top performer in my classes, I had never really mastered the art of studying. Somewhere along the way, I came to believe that highlighting was the key to learning—so, my thinking went, the more I highlighted, the more I would learn. My biology and chemistry textbooks turned into veritable neon lights as I dutifully highlighted practically every word in my misguided belief I could absorb all the material. My early test scores revealed the error of my ways. I called Momma on the phone in tears, telling her, "They're so much smarter than me. I'm not going to make it." In her usual stern way, she told me to buck up and work harder, that I was there for a reason and could succeed if I tried. Slowly, I started to learn how to study—writing up short summaries of my readings and creating index flash cards for large memorization tasks. And I made great new friends, like Sahil and Jenny, who shared their study techniques with me. Eventually, I weaned myself off my highlighter addiction.

Beyond my intensive premed course load, I wanted to take some classes for fun, something to stretch my brain and give me a break from rote memorization. The first of these was Introductory Anthropology—from there, I kept going deeper. I loved seeing the myriad ways humans celebrated life through foods, customs, language, and art. In my Medical Anthropology course, my fascination with other cultures merged with my love of medicine. Suddenly, though, it wasn't just being introduced to ancient civilizations or faraway tribes, I was being *reintroduced* to my own culture, my own story, my own body. What was the "best" form of medicine? Was there *one*, or were there multiple? Were we missing the social and psychological benefits inherent in traditional modes of healing in our modern Western approach? What is *health*? Are measures of health also dependent on the cultural lens through which health is viewed? What is disability, and what does it mean to be identified as disabled? I learned that the meaning of the disabled identity could shift from culture to culture. In some cases the disabled were viewed as weak and to be discarded or left to die, whereas for others it was a mark of favor by the gods or a sign of special healing powers or other gifts. *Where does that leave me?*

Before I knew it, I had accumulated enough courses in anthropology that I was able to add it as a second major to my primary major in biology. Through it all, as I worked toward the course requirements for medical school and prepped for the MCAT exam, I felt something was missing. While I still wanted to pursue my dream of becoming a surgeon, I missed the thrill of research I had experienced during all those science fairs, and had fallen in love with the fieldwork aspect of my anthropology courses. I wondered if this was what it meant to become an adult, choosing a left or a right at a fork in the road, knowing you were forever saying goodbye to one life path while you pursued the other. That was when I discovered ethnobotany.

In a short two-credit course in Tropical Ecology, Dr. Larry Wilson mentioned the field of study I would one day make my own. A lively man with brownish-gray hair and an avid love of amphibians, Dr. Wilson explained the bridge between humans and nature—and how human knowledge of plants and animals, passed down from generation to generation, is what led to the development of various cultivated foods, medicines, clothing, and tools. When I read one of the assigned books for the class—*Tales of a Shaman's Apprentice* by Dr. Mark Plotkin, which recounts his travels in the Amazon and his training under a shaman—something clicked in the back of my mind. This was a story of origins—the origins of our modern pharmacy that sprouted from the wisdom and practices of people intimately acquainted with nature's bountiful resources.

I found out that Dr. Wilson took a small group of students to the research camp at Explorama Lodges in Peru for a spring break trip nearly every year. The one-week tour exposed them to different habitats of the forest as they looked for wildlife and plants, spanning the flooded forests to high into the canopy. I really wanted to go, but it was out of my price range. I asked him if there were any research trips I might be able to do for less money. Amazingly, he said there were opportunities to work at the camp as a research intern with the gardens for longer periods of time and at a heavily discounted rate—just thirty-five dollars a day for room and board.

With Daddy's encouragement and help on gathering the funds, I began planning my trip for that summer: I would fly to Peru in the summer of 1999, the new millennium just around the corner. This was my chance to explore that something I felt was missing in my life and undertake a research project of my own design. Maybe this wasn't a fork in the road. Maybe I could honor both my passions and blaze my own trail up the middle.

CHAPTER 2

Welcome to the Jungle

The most important affair in life is the
choice of a calling; chance decides it.

BLAISE PASCAL, *PENSÉES*, 1660

S eated on a rickety bench propped up against a tree trunk, I waited
for the medicine man.

A large bush offered a shady respite from the sweltering midday sun,
its branches adorned with dark glistening leaves and spiky fire-engine-
red fruit pods the size of a chicken egg. It reminded me of a Christmas
tree laden with holiday ornaments. I admired the fingerlike fans of leaves
sprouting from the *Cecropia* trees of the Urticaceae family, their open
treetops resembling candlesticks. A brown lump appeared to hang off
one of them, moving ever so slowly, its long claws creeping toward the
main trunk—a sloth. Three blue morpho butterflies swooped across the
path nearby, their large metallic wings glistening in the light with each
flutter through the air. A chorus of birds pierced the steady hum of insects hovering around me. The diversity of life was a delight to every one
of my senses, surrounding me from the dark blue skies to the muddy
earth caked beneath my boots.

The Amazon contains the most biodiversity on earth. It is home to more than 1,500 bird species—that's nearly 1,000 more bird species than are found in all of Europe! There are more than 2,500 species of fish, 1,400 species of mammals—including the prowling jungle cats, the jaguar and the ocelot—as well as the friendly capybara (the largest rodent in the world), the South American tapir, monkeys of many kinds, and anteaters that roam the forest floor. There are more than 1,500 species of amphibians and inestimable numbers of insects, and the forests are full of many species that have yet to be described by science.

The medicine man was named Don Antonio. I sat on the bench while he fetched something from his palm-thatched hut in the center of his garden. As I waited, I tried to knock off some of the mud glommed onto my hiking boots from our expedition into the jungle that morning. No such luck, still too wet. My beat-up backpack was lying beside me on the bench, stuffed with gear and tools—camera equipment, clippers, a first-aid kit, a water bottle, iodine tablets, a copy of Al Gentry's *Field Guide to the Families and Genera of Woody Plants of Northwest South America*, a notebook, and a pen.

I arrived at this remote part of the Peruvian Amazon just a week prior—plane from Atlanta to Lima, then Lima to Iquitos, where I waited a couple of days for a motorboat to take me upriver. Iquitos is a remote jungle town inaccessible by road but still the sixth most populated city in Peru. During the rubber boom of the late 1800s and early 1900s, it was a major center of production. Today, its economy is based on the export of various forest resources, ranging from timber to fish, and supporting a stream of tourists that flock to this gateway of the Peruvian Amazon.

The boat traveled up the Amazon and then moved northwest into the Napo River, nearly reaching the Ecuadorean border before arriving at Explorama Lodges situated on the Sucusari River. It was June of 1999 and I was a rising senior in college. I'd just turned twenty-one, and it was my first trip to South America. I'd taken plenty of family road trips—and even a class trip to Europe in high school—but going it alone in such a wild place was a totally new experience. As a research intern with Explorama Lodges, I was lending a hand in the ethnobotanical gardens run by Don Antonio and his son, Gilmer.

Armed with a letter of introduction from Dr. Wilson and basic Spanish language skills from high school classes along with bits of medical Spanish picked up while volunteering in the ER, I settled into camp life. Don Antonio's garden grew only plants traditionally used for food, medicine, and tools. Along with tending to the garden, Don Antonio and his son led tours for the tourists who came to the Sucusari lodge to explore both the gardens and the canopy walkway situated deeper in the forest, near the research base.

When Don Antonio ambled back to me, he reached up and grabbed one of the spiky pods that dangled above our heads. Don Antonio was a few inches shorter than my five-foot-six-inch frame, with dark eyes, charcoal-black hair, and broad shoulders. Though his hands were calloused from his daily work with plants, he was no simple gardener. He was also an Ayahuasquero shaman who used plants to heal patients in surrounding villages. The garden gig was just his day job. Don Antonio told me that this plant was called achiote; I dutifully found it in the pages of my field guide. It was *Bixa orellana* in the Bixaceae family. As would become a habit between us, he spoke in a lilting voice, patiently explaining a plant's unique qualities, stopping to repeat words in Spanish I didn't understand so that I could look them up later. I was engrossed, scribbling my notes as fast as possible, eyebrows furrowed in a wavering mixture of concentration and confusion.

He broke open one of the pods along its seams, revealing waxy red seeds, about half the size of a cranberry, similar in color, but with a tougher texture. He began smashing the seeds inside the split fruit pod with his finger, rubbing them around until a thick red paste was created. Intrigued, I took detailed notes on this process next to a sketch of the fruit. Next, he leaned close and started to apply the paste to my lips and around the sides of my mouth. As he worked, my mind raced. Could this be a remedy for viral cold sores? Chapped lips? Something else?

He carefully took the bag he had gathered from his hut and pulled something out that flashed in one of the columns of light falling through the canopy. It was a small mirror, which he held up to my face. I saw that the red paste was smeared across my mouth and teeth.

His sly grin broke into a broad smile and he pointed at me.

"Lipstick!" he said before his belly laugh nearly convulsed him to the forest floor. I began laughing, too. I looked like a clown.

PEOPLE HAVE USED PLANTS as their primary source of medicine since we first found our way out of the African savannas, walking on two legs. While hiking in the Alps in 1991, two lucky people stumbled upon the frozen remains of the 5,300-year-old "Iceman" Ötzi, along with hunting tools, berries, and polypore fungi. The birch polypore produces agaric acid, a strong purgative that Ötzi may have used to treat whipworm—which scientists found in his gut. Some of the earliest written records of medicinal plant use date back to ancient Egypt circa 1550 BCE, and in one scroll that is over sixty-five feet long, there are some seven hundred plant formulas and remedies for treating various ailments, including chronic wounds and skin diseases. Another early record linked to an ancient medical tradition was written in China around 200 BCE, known as the Drug Treatise of the Divine Countryman (Shennong Ben Cao Jing). It contains an inventory of 365 drugs as well as each plant's geographic origin, collection time, therapeutic properties, preparation, and dosing.

The Indigenous people of the Amazon Basin have also used natural resources to effectively manage health and treat disease for millennia. Which ingredients to use and when to harvest, how to prepare them and determine the dose, and when to prescribe it is part of a vast body of knowledge passed down from shaman to apprentice through oral traditions. There are an estimated four hundred tribal groups living in the Amazon—each with its own distinct language, territory, culture, worldview, and system of medicine, and their medical traditions are as varied as the people themselves. With each tribe lost to the influences of the dominant Western culture and its market economy encroaching into the forest, with each language lost, and with each shaman who dies without an apprentice to carry on the vast amount of knowledge, history, and traditions of their people, the world forever loses the equivalent of libraries full of wisdom. As a medicine man, Don Antonio relied solely on the resources the forest provided and what he grew and harvested in his garden to treat his patients. As the custom goes, he had learned about these remedies from his apprenticeship as a young man, and now he had begun the process of transmitting his hard-earned knowledge to his own son.

But just how valuable was this knowledge? How effective are these forms of medicine? Could their use be explained by the simple principles of placebo? Were these just old wives' tales based in legend and stories but without scientific merit? These were some of the questions I had wondered about before embarking on my journey to South America. While I was eager to learn more, I was also skeptical. I hoped to find some answers.

In my desire to be prepared for anything on the trip, I brought everything I could imagine I might need—lightweight T-shirts and hiking pants, research gear, botany and language books, and, of course, a full suitcase of medical supplies (e.g., assorted bandages, Band-Aids, antibiotic creams, oral rehydration packets, antidiarrheal pills, Benadryl, rubbing alcohol, suture kits, malaria pills, and Silvadene burn cream) that would have been enough to stock a field clinic.

My room was small and spare. Besides the twin-size bed, a simple mattress on a wooden platform with mosquito netting hanging from above, my room contained a chair and small table, where a large bowl and a pitcher of water sat to use for washing up. There was no electricity, just an old-timey oil lamp with a wick. A door and a wall faced the outer hallway, but the other side was just a half-wall, open to the moist jungle air.

Each day before dressing, I went through the ritual of carefully shaking out my clothes and shoes to ensure that no scorpions or other stinging critters had snuggled into their folds and crevices. I followed this with generous applications of sunscreen and bug spray. If there was anything more relentless than the sun, it was the mosquitoes. This was a malaria-endemic zone, with cases common among the local villages, so I didn't want to take any chances. The chloroquine pills I took to prevent malaria gave me intensely vivid dreams at night—featuring giant serpents, jaguars, and creatures of the dark that would bring me to a startling state of wakefulness as I disentangled my mind from the nightmare. Despite the liberal use of my deet-infused bug spray, my arms and legs were still spotted with angry red welts from undeterred mosquitoes.

I wasn't the lone American visiting the research camp. Another student, a petite girl named Jane from Washington State, was also volunteering in the camp garden. Sporting long, flowing shirts, perfumed in patchouli, and reluctant to bathe, Jane had blonde dreadlocks that cascaded down the middle of her back. We were a study in opposites. I was the premed student wearing military-issue camo pants, a plain white T-shirt, a utility belt for my canteen and bowie knife, and a barbershop buzz cut around which I tied a bandanna to keep my head cool on the long, hot days.

Jane was nice, if ethereal—a strong breeze seemed capable of whisking her away—and she did her best to introduce me to the local staff at the camp and show me the jungle path that ran through the forest from one camp to another. The path had been cleared years ago and was meticulously well kept by the staff and guides. Carelessness in the thick underbrush could lead to an accidental run-in with one of the many deadly creatures: my chief concern was keeping an eye out for the dreaded fer-de-lance, a venomous pit viper known to leap out from its coils to strike, especially during mating season.

Despite my overzealous preparation for any possible emergency, I landed myself in trouble within the first week. Jane insisted on showing me a shortcut between the research camp and Don Antonio's garden. I'm adventurous—but not stupid (or so I thought). If I'm new to an area and without compass, map, or GPS, I'll stick with the trail. But Jane insisted the shortcut was a beautiful walk, and as she was better acquainted with the landscape, I deferred to her judgment.

"We'll be there in no time," she said confidently.

While it was certainly a beautiful part of the forest, short it was not. It was also not an easy hike, requiring us to scamper over fallen logs, dodging thorn-laden tree trunks, and travel up and down hills and across small and slippery streams. This was especially challenging with my prosthetic leg, and the entire time, I worried we would interrupt a pair of fer-de-lance in the middle of mating and become the next fertilizer for the rain-forest floor. Jane was unflappable, practically floating along the path.

At an intersection with a small stream, she dipped her cupped hand into the cool water.

"You really shouldn't," I cautioned. "Here, take some from my canteen."

"No," she said, drinking from the stream. "It's pure."

I had filled my canteen with pre-boiled water at the camp and carried iodine tablets with me to purify water in a pinch if needed. She said she drank the natural wild water all the time. Alarm bells clanged in my premed head: there are helminths, bacteria, and all sorts of parasites that will destroy a newcomer's gut!

When we finally arrived at the garden, I was exhausted, dehydrated, limping, and more than a little pissed off. My stump was rubbed raw, bathed in salty sweat from the hot hike, bright red from the irritation. I was able to rest for a bit but then had to begin the walk back to camp for dinner before it got dark. This time I took the marked trail, an easy thirty-minute walk.

From the moment I woke up the next day, I knew something was wrong. With an infection, it's either the odor—that sweet, cloying, putrid smell Momma caught a whiff of after my first surgery—or deep, persistent aching pain. I disentangled the mosquito netting from my bed and sat up on the edge, shivering from the chill of the cool, humid morning air.

Even though I carefully washed my stump the evening before, slathering zinc-oxide baby rash ointment to soothe the inflamed skin, the telltale signs of swelling instilled in me a sense of dread. I twisted my leg to take a closer look and then groaned. An angry volcano emerged from the tissue of my calf muscle, jutting out from the sea of white paste. At the top of the mound, the skin was taut, discolored a bluish black. A boil had formed right at one of the major contact points where the skin under my knee rubbed against the tough plastic socket of the prosthetic leg.

Human skin is coated with bacteria all the time, and given the right opportunity and conditions for growth, even otherwise harmless commensal (normal and beneficial) microbes can become problematic. The primary role of skin is to act as a barrier to protect our organs and blood from attack by foreign substances, whether biological or chemical. The hot, wet environment of the rain forest combined with the damage I had done to my skin barrier at the irritated site presented the perfect place for bacteria to converge and multiply. This was not the first time I had been confronted with this problem, but it was the first time I had dealt with it alone, separated from a modern medical facility by days of riverboat travel.

As a young girl, I'd frequently suffered from similar boils that came after days of running around and playing outdoors, the friction of the stump socks rubbing the skin raw. After slicing the boil to drain and clean the wound, Daddy would hold me as I sobbed through the pain. "If I could cut off my own leg and give it to you, I would," he'd say.

I gently prodded the infection with the tips of my fingers. It was hot to the touch and painful. The head of the boil had not yet matured. I would have to wait until it came to a head before I could lance it. Until then, I'd be bedridden.

No crutches. No backup wheelchair. Nothing. Without my prosthetic, I was immobilized.

I cried in angry frustration, mad at myself for not being more careful, mad at this shitty hand of cards fate had dealt me with this crippled body, kicking myself for listening to ethereal Jane. This was not how things were supposed to work out for me on this trip. I had so much to do and experience, and only a month remaining. Again, my body failed me. There was nothing to do but wait. So I did. I read books on the botany of the region, but every page I turned felt like another real-world opportunity out there in the great humming jungle that I was missing.

After two days, the boil matured. Knowing exactly what to do from the hours I'd spent observing physicians and nurses while volunteering in the ER, I first used the Betadine from my medical kit to sterilize the skin. My years of studying bacteria in the hospital clinical microbiology laboratory had taught me a lot about the importance of sterilization, so, taking the glass cover off the oil lamp that lit my room, I dipped the tip of my bowie knife in alcohol, then set it aflame.

I took a deep breath and made the incision. Pus and blood finally gushed out, and I gritted my teeth through the pain. I used my gloved fingers to squeeze the flesh in an effort to force out as much of the material from the wound as I could. After wiping away the viscous mess, I used a packet of sterile saline and more Betadine to rinse out the wound, then dressed it with clean gauze and topical antibiotic ointment.

Finally! It was time for . . . more waiting. It was like I was back to being a kid again post-op, except now I was stuck indoors in one of the most exciting regions in the entire world.

After several days in bed and many more pages turned (apparently, the intrepid nineteenth-century naturalist Alfred Russel Wallace developed a profound understanding of the interdependence of palms and people of the Amazon during his exploration of the Rio Negro), the wound healed enough for me to wear my prosthetic leg. I asked the staff to radio in a request for me to be relocated to the main tourist lodge, adjacent to the gardens, rather than the research station deeper in the forest. I wanted to avoid additional skin flares by keeping long, sweaty walks at a minimum. They graciously gave me a spot at the other camp, and things were much easier thereafter.

Once I settled into my new accommodations, I heard from the staff that Jane, too, had met her own misfortune. Not long after my leg incident, Jane traveled to Iquitos to get medical treatment after developing a bad case of giardiasis from consuming food or "wild" water contaminated with animal or human feces infected with *Giardia* species. *Giardia* is a tiny parasite that can cause serious abdominal cramps, chronic diarrhea, weight loss, and dehydration, requiring a course of antiparasitic drugs. I later learned that she recovered and returned to the United States the following week.

I was lucky not to have required hospitalization and IV antibiotics for my infection. I knew the risks of coming here. The hot, wet environment combined with frequent activity would present the same threats of infection I'd experienced as a child in Florida, only this time I'd need to rely on my own medical skills because access to a hospital would be difficult. This would not be the last time my stump became infected.

In Don Antonio's ethnobotanical garden, I imagined myself a visitor to Eden—surrounded by fragrant flowers and fruit, and plants of every shape and size, all organized carefully in quadrants, with neat rows of medicinal herbs, clusters of shrubs, and various useful trees interspersed. These were plants used to make medicines, tools, foods, art, and more. This small plot of land held everything needed for survival—and it all just grew there!

For a student fascinated with medicinal plants, it was a wonderland.

For Don Antonio, it was his pharmacy, the neat rows forming stocked shelves from which he could find anything he needed.

There were papaya trees (*Carica papaya*, Caricaceae), whose green, unripe fruits decorated their trunks. This unique trait of growing flowers and fruits on the trunk is known as cauliflory—and it allows trees to be pollinated or their seeds dispersed by animals that climb up and down the trunk. Don Antonio taught me that the white latex that can be extracted from papaya trees was used locally to treat intestinal parasites and as an abortifacient.

Nearby, the grasslike piripiri herb (*Cyperus articulatus*, Cyperaceae) could be quickly harvested to treat snakebites, digestive disorders, or even fever, flu, and fright disorders. There was also the tree known as bellaco caspi (*Himatanthus sucuuba*, Apocynaceae); the latex that it bears is topically applied to wounds and to treat lumbar pain. There was toé, or angel's trumpet (*Brugmansia suaveolens*, Solanaceae), whose long pink flowers popped out from its shrubby branches. An alkaloid from this plant, scopolamine, is used as a medicine in Western practices as the main ingredient in skin patches applied to treat motion sickness and postoperative nausea. The leaves and flowers of angel's trumpet are macerated in water and taken orally as a hallucinogen by healers, or curanderos, to communicate with the spirit realm.

One day in the garden, Don Antonio sliced his machete across the trunk of the sangre de drago, or dragon's blood tree (*Croton lechleri*, Euphorbiaceae). From the wound, a dark red resin emerged, almost like blood seeping out from a laceration of human flesh. He took his finger and ran it along the bleeding slash, then took my hand and smeared the sticky resin in my palm. At first I thought this might be another practical joke ("Lipstick!"), but he insisted with a furrowed brow that I use my other finger to rub the resin in my palm. As I did, the substance turned from deep crimson to a pale gray. The compounds in the dragon's blood seeds had reacted with the oils of my skin.

"This is how you know if you have the right medicine," he said. "In the markets, they sell red dye to tourists in glass bottles, not real medicine from the plant. This is how you know."

The resin, he explained, treated minor skin infections. He took some from my palm and rubbed it onto my arm where scabs had grown over mosquito bites that I'd scratched. "This will heal your wounds," he told me. He went on to describe the internal use of the resin for the treatment of diarrhea and hemorrhaging after childbirth. "Strong medicine," he explained.

I took dutiful notes, but I remained skeptical. Could this color-changing plant he had plucked from a tree really be strong? Plants are interesting . . . but how useful could they be, *really*, when faced with treating a serious, even deadly, disease? Mosquito bites and scratches are one thing, but internal bleeding and diarrhea are another.

Years later, I would find out—and so, too, would the rest of the world. In 2012, the resin of this dragon's blood species received the rare and hard-won approval of the FDA as a botanical drug, signifying that it passed the hurdles of rigorous manufacturing controls and clinical trials to demonstrate safety and efficacy. The most effective version of the drug was not a single compound, but a mixture of compounds found in the resin. Today, it is known as crofelemer and is marketed as Mytesi—an oral prescription drug for the treatment of noninfectious diarrhea without the risks of causing constipation, as happens with opiates. Don Antonio was right—it was a powerful medicine.

As we continued through the garden, I pointed at another tree. "What about this one?" At almost seventy feet tall with large buttress roots at its base, the tree grew leaves elliptic in shape and pale green in color with a yellow-lined venation pattern. The trunk was pale brownish gray, somewhat mottled, yet smooth to the touch. Amazingly, it began its life as a climbing vine that clung to a host tree, eventually killing it and taking on a final form as a mature tree. I'd seen Don Antonio cut the bark of this tree before, collecting white latex into a small bottle. From my field guide, I learned that its scientific name was *Ficus insipida*—a fig tree in the Moraceae family.

Don Antonio drew his knife and sliced into the bark, and white fig "milk" emerged. "This one is ojé," he explained. "The milk should be mixed with fruit juice and given to children to purge them of the worms in their bellies. But you have to be careful. Only a curandero should do this, because if you give too much, it can poison the child."

As a curandero, Don Antonio was taught to serve as both healer and pharmacist—he prepared the drugs and he prescribed them. The years of knowledge and training required to recognize classic determinants used in dosing of pharmaceutical drugs also applied here—the weight and health status of the patient must be considered when prescribing a therapeutic intervention. The line between poison and medicine often comes down to two simple principles: dose and intent.

Don Antonio taught me about many different species within his magical garden. What never failed to amaze me was that despite his lack of education beyond the rural village school in his early childhood, he possessed a deep and sophisticated understanding of human physiology and pharmacology, even psychology. Moreover, this expertise all derived from one thing: nature.

ONE DAY WITHOUT WARNING, Don Antonio asked about the muscle and phantom leg pains I suffered from. He noted that I seemed achy, moving more slowly than usual as we worked in the garden, weeding and harvesting. I'd been struggling with major life decisions that left me distracted and sluggish—not that Don Antonio knew anything about them. Ever since I started volunteering at my local hospital, I wanted a career in medicine as a practicing physician, a surgeon. But as I gained more and more exposure to the people of the rain forest, learned more about the gaps in public health, and pondered how Western and traditional medicine intersected (or didn't), I began having doubts. And these doubts were rocking the foundation of my grasp on who I was and what I was meant to do with my life.

"Would you like me to heal your spirit today?" Don Antonio had stopped tending to a plant in his garden and was now facing me, his expression serious, asking if I would like for him to treat me. *Spirit?* What did that even mean? This was certainly unlike the vacation Bible school lessons on spirituality I'd had as a kid. Was this some sort of traditional psychiatric intervention?

I found this interesting, that his first instinct was not to directly heal the aching weak muscles of my right thigh or the shooting nerve pains that plagued me. He wanted to heal my spirit—my full sense of being. Maybe he sensed the dilemmas I faced just by learning my moods.

As an Ayahuasquero shaman, Don Antonio sometimes used a mixture of medicinal plants to yield a thick hallucinogenic brew. I knew from reading books that this yielded powerful effects on its consumer, first causing vomiting and diarrhea, followed by an experience of intense colorful visions. Artists painted images from these visions, filled with forest animals, often attributed as "spirit animals" in the local cosmology, such as jaguars and long, fat, slithering anacondas, represented in bright colors and embedded in the forest ecosystem with various trees, flowers, vines, and sometimes even people.

Don Antonio began his training as a curandero at a very young age—just nine years old. When we first talked about his training, I was shocked to learn that his grandparents, who raised him in a small village on the Napo River from the age of four, left him on his own in the forest for a month. His grandfather took him deep into the forest until they came upon a great ceiba tree. He told Antonio that he would need to remain there in solitude, not speaking to anyone. He was instructed to drill a hole into the tree and tie a gourd to the hole and leave it for eight days. After this he should eat the gelatinous sap that pooled in the gourd. During this month, he was to fast, without sugar or salt, just eating a little dried fish and the sap of this tree.

"As a child, my grandfather told me stories of the forest spirits. In those days that I was left alone in the jungle, I saw things I'd never seen before, and I was scared. I tried to run away and was beaten for it. They made me promise not to run away anymore," he explained as I nodded in sympathy.

"It is hard to be a shaman," he sighed.

As I learned more about his way of medicine, I saw how this childhood experience, as well as the later ones he'd had on the use of ayahuasca as a means of communicating with those same forest spirits he'd been terrified of as a child, were critical to his development as a healer. For him, ayahuasca was never a substance used for recreation. It wasn't employed for a Goop-style retreat in LA or a wild night of "finding himself" in a Brooklyn brownstone with copies of *Be Here Now* on the coffee table. Instead, for him, ayahuasca was a true entheogen—a natural psychoactive substance used to achieve a spiritual experience, always in a sacred context. In his case, this experience was aimed at supporting his development as a healer.

Every single detail—from the harvest of the plant ingredients to the preparation of the thick brew, which simmers for hours over a campfire; to the consumption of the beverage, which is embedded within a ritual of whistling and the shaking of a leaf bundle (known as shakapa or chakapa in Quechua); to the spiritual journey that ensues—is sacred. When confronted with patients he wasn't able to diagnose, or needing help determining which plant medicine would be most beneficial, Don Antonio took ayahuasca as a means of communicating with the forest spirits. Ayahuasca was his teacher, and he even greeted it as such during the ritual of preparation and consumption. He explained that in his visions, he traveled within his mind to the place in his garden or the forest where the healing plant grew; along the way, he communicated with the animals and plants of the forest. Ayahuasca was a key component of his diagnostic and prescriptive process.

The ayahuasca vine (*Banisteriopsis caapi*, Malpighiaceae)—also called the vine of the soul, yagé, or caapi—is just one of several ingredients found in the ayahuasca brew. It grows up to thirty meters in length, winding among other plants in the forest for support. The age of the vine and its point of optimal pharmacologic potential are measured by curanderos based on its girth.

Every ayahuasca curandero uses a special recipe to create their brew, which is composed of different species that they learned about during their apprenticeship or through experimentation. These are often held secret, but *Psychotria viridis* (from the coffee family, Rubiaceae), locally known as chacruna in the Quechua language, has been a widely reported ingredient used as the source of DMT (dimethyltryptamine, a natural hallucinogenic compound) in many ayahuasca recipes. Normally, DMT breaks down in the gut, but combining it with the harmala alkaloids in ayahuasca that act as MAOIs (monoamine oxidase inhibitors, a type of antidepressant) prevents rapid breakdown and allows the DMT to reach the circulation and central nervous systems. When ayahuasca is orally ingested with MAOIs, the effects of vivid visions and the sense of euphoria can last three or more hours.

Ayahuasca tourist camps are scattered around Iquitos as part of a new tourism trend that has emerged in the Amazon over the past twenty years or so. This concerns me. As a sacred plant ritual held in respect for millennia by healers who dedicate their lives to its use for healing, it should not be used as a tourist gimmick. To do so is to divorce it from its original cultural context and value. Moreover, MAOIs should never be combined with prescription or over-the-counter antidepressants, cough remedies, and painkillers. The MAOIs present in the brew can cause some serious drug interactions if taken with pharmaceuticals, and tourists seeking resolution of mental health issues may be at greatest risk. Combining MAOIs with SSRIs (selective serotonin reuptake inhibitors, drugs that increase levels of serotonin in the brain as antidepressants are designed to do) is sometimes lethal.

As with many complex remedies, I've spent hours pondering how people discovered and refined them in the first place. How did Don Antonio's ancestors determine that mixing certain ingredients of plants not normally used as food could yield such an effect? Was it through observations of animals using the plants? Many different species have been found to self-medicate—chimpanzees chew the bitter pith of *Vernonia amygdalina* (Asteraceae) to remove intestinal parasites—and the idea that human discovery has been fostered in part by observations of animal behavior has logical basis. Or was it discovered by chance, confusion with another edible resource? Steady experimentation with environmental resources? No matter what the path to discovery, one thing is certain: humans have engaged in experiments with environmental resources and made observations of nature since time immemorial. Through the sharing and passing down of knowledge of useful species from generation to generation, in either oral or written form, we have all benefited from this collective knowledge.

THE SPIRITS ARE STRONGEST on Tuesday and Friday," Don Antonio explained.

I had accepted with gratitude his offer of healing, intrigued by the gesture and curious to experience this form of healing.

And, as it happened, it was a Friday.

Don Antonio did not offer the ayahuasca brew for me to consume. Instead, he told me that he had already conferred with his teacher, the spirit vine yagé, about my health issues. I was to meet him later that day at the edge of the garden.

When I arrived, he stood in the shade, wearing a long black robe I'd never seen before. His face stoic, lacking the sly grin I'd become accustomed to, he held a bushy bundle of leaves made of a *Pariana* species (from the grass family, Poaceae). From our prior conversations about the shakapa bundle, I knew this was special, used only for healing as a tool to focus communication with the forest spirits.

A small handmade stool stood in front of him, and he motioned for me to sit. I crossed my ankles, sat up straight, and looked forward. He took my hands and placed them palm-side up on my thighs, then took a bowl filled with water and crushed mint leaves, freshly harvested from his garden, and began to rub it on my arms and face. He then took my head in his palms and held it, applied pressure for a moment, then poured the remaining herb liquid into my hair. The water was cool and fragrant, tickling my neck as it dripped down. Unsure of what would come next, I was also eager to continue.

Behind me, he placed one hand on my shoulder and the other on my head, then began a speech to the plant spirits asking them to come to help me. He called on the spirits of women and all of the good spirits of the forest to help me with my future—to give me intelligence and to guard and protect me in my path toward good medicine. I closed my eyes.

He began to whistle softly, moving his hand up and down, creating a smooth rhythm that I could sense not by sight but by sound as the shakapa bundle rattled. A tingle zipped up my spine. His whistle grew louder. He moved around me in a circle, shaking the shakapa. I let him and the rhythm surround me.

As time went on and the sounds continued to swirl around me, my mind began to wander, dreamlike. Eyes still closed, I inhaled the muggy, late-afternoon air, smelling the garden herbs that perfumed my body. Beneath the shakapa rattle and Don Antonio's lyrical whistle, I began to hear another chorus, the forest, singing a medley of buzzing insects and frogs. Birds chirped along with Don Antonio, and I could hear movements of monkeys and other mammals in the network of branches and vines that formed the forest canopy swaying above us. Suddenly, I didn't feel apart from the jungle. I belonged to it.

Don Antonio began to sing. I didn't know if they were words of an old language or sounds of the moment. But they sounded right. He paused his song and motion at one point and made a loud sucking noise directed at my head. He started on the right side, sucking at the air as one would with a clogged straw, moved to the center, and finally the left side. He resumed his whistle song and shook the shakapa bundle, lightly tapping the crown of my head three times, followed by three more sucking noises, symbolically drawing out the undesired elements of my spirit. He ended the ritual with one hand on the front of my forehead, the other at the back of my skull, applying the same light pressure that he began the ceremony with.

After it was over, we walked back to the center of the garden and he explained that the spirits would stay strong in me but that it was very important that I abstain from sex for six months. He had collected some leaves from the guayusa tree (*Ilex guayusa*, Aquifoliaceae) during our walk, and the dark green waxy topside of each leaf glistened in the sunlight; he crushed them in a basin of water and carried it inside the hut. He put up a blanket in the doorway of his dirt-floored palm-thatch hut for privacy and told me to bathe with these herbs to finalize the ceremony. As I was washing up, my body covered in bits of leaves, I stopped.

How had I gotten here—in this moment, in this place?

I grew up in a rural area of southwest Florida and was raised in the church—first as a Southern Baptist and then as a Methodist. I knew the value of prayer and meditation and the impact a spiritual practice can have on a person, especially in fostering a sense of connectivity. Having undergone so many surgeries, I was also familiar with the buzzed sense of relaxation right before going under general anesthesia, or the morphine-drip bliss during post-op recovery. But this was different; it was unlike anything I'd experienced before. I wasn't buzzed. I wasn't euphoric. I also wasn't in pain as I had been earlier that day. For the first time in months, maybe years, I felt whole. I felt grounded. United with things greater than me, greater than my body, greater than my spirit. A part of the forest. A part of the world. Integrated. Unified. Healed.

Don Antonio had achieved this feeling without drugs, just a fragrant herb bath, the power of human compassion, ritual, and song. I couldn't put my finger on it at the time, but I was beginning to learn not to dismiss the things I didn't understand. I was learning to ask deeper questions. Maybe medicine was about more than pharmacology and surgery.

CHAPTER 3

Worms in the Belly

Our imagination is struck only by what is great; but the lover of
natural philosophy should reflect equally on little things.

ALEXANDER VON HUMBOLDT, *PERSONAL NARRATIVE OF A JOURNEY
TO THE EQUINOCTIAL REGIONS OF THE NEW CONTINENT*, C. 1814

The peke-peke motorboat burped its way up the Napo River toward
the Ecuadorean border as I leaned against the port side, my fingers
trailing in the water. I watched the shoreline and waved to the children
splashing in the shallows along the banks of the villages we passed.

I was back in the Amazon only a few months after my first trip. I was
hooked. It was December, the wet season, and I was riding on one of the
main arteries of the Amazon Basin. While I spent the entire fall semester
on campus in Atlanta furiously applying to med schools, I'd begun plot-
ting to find a way to return to the rain forest almost immediately after
stepping off the tarmac. And with a bit of ingenuity, I found one.

I applied for and won an undergraduate grant for $3,000 to build
upon the preliminary data gathered over the summer, investigating the
ways that modern Western medicine and traditional medicine affected
child health in the mestizo communities situated along the upper Napo
River. If Don Antonio was any indication, traditional medicine was a

rich source of knowledge. At the same time, I knew that a lot of infectious diseases cropped up in the Peruvian Amazon and Western medicine could play a role. The scientific literature of the time included many studies that reported on the traditional uses of plants for health and raised concerns about the loss of traditional knowledge, but few examined this intersection of Western and traditional medicine, or how the displacement of medical traditions influenced population health. I wanted to learn more about this crossroads in medicine.

As the largest tropical rain forest in the world, the Amazon makes up 40 percent of the landmass of South America, covering nearly three million square miles of land and interweaving networks of tributaries, rivers, and streams. The main water highway of the jungle—the Amazon River—begins in the Andes Mountains to the west, traveling roughly four thousand miles eastward into Brazil, where it eventually drains into the Atlantic Ocean.

As we rode down the basin, a barge towing rows of massive tree trunks approached in the distance. Centuries of old forest growth were being cut down in an instant. The commercial logging enterprise was big business, and I pondered how much it helped or hurt the local people. Multinational corporations were often behind these large-scale extractions, and while there might have been some local income as a result, the people of the region were left with forests bereft of the foods and medicines they'd survived on for centuries. The rainy season made it easier for loggers to penetrate deeper into the forest, floating felled trees downriver toward Iquitos.

I was on my way to meet with a local sanitario—a government healthcare worker who tended to patients from the surrounding ribereño communities. *Ribereño* refers to the "river people," communities along main rivers no longer affiliated with specific tribes or ethnic groups and who have been westernized to varying degrees, speaking Spanish and practicing Catholicism. They are also known as mestizo communities, having mixed Indigenous and European ancestry. During the era of the rubber boom, Christian missionaries rushed into the jungle, influencing local cultural and religious practices. Unlike the Indigenous Yagua and Maijuna communities, who more commonly dressed in traditional clothing made of local plant materials like palm fibers and whose villages were hidden deeper down small tributaries, both young and old ribereños dressed in frayed hand-me-down factory-sewn clothes, often the imported donations of Western consumers resold at markets. Typical ribereño villages featured a government-constructed school assembled from concrete blocks, whereas the villagers continued to live in handmade open-air stilt-elevated huts made of local timber, their roofs thatched with handwoven palm leaves.

I found the sanitario Señor Vidal in his health clinic, elevated on stilts with a large open platform and a small private section for medical exams. A throng of patients waited their turn to be seen, sheltered under the cool shade of the palm-thatched roof—young children playing games, pregnant women resting in palm-fiber hammocks, a few young men standing around with minor injuries and cuts requiring sutures. A young sloth—likely a pet of one of the children—clung to one of the corner poles of the hut, slowly munching on *Cecropia* leaves as it observed the scene.

Beyond the waiting area, there was a single desk with two chairs. The wooden top was well worn, but clean and organized. In the center was a large ledger, where Señor Vidal wrote his case notes, neatly lined rows of information on each patient—age, gender, village, ailment, and treatment. On the right side of the desk sat a single microscope. As there was no power in the clinic hut, Señor Vidal used the reflective light of the sun hitting the microscope's mirror when examining slides. A stack of finger-prick blood smears on glass slides sat in a box next to the microscope—this was how he diagnosed cases of malaria, an endemic, chronic problem.

I introduced myself, and he agreed to speak with me over lunch. In the meantime, I waited with the others as he worked through his caseload. He called over a young mother with her toddler. She couldn't have been older than seventeen. A teenager unsure of what to do for her sick child—my heart ached for her. The little boy was plump and well fed, but clearly sick, coughing and feverish with a respiratory infection. Señor Vidal gently prodded the boy's neck and listened to his chest, nodding as the mother explained the symptoms, and recorded handwritten notes in his ledger. He spoke with her in hushed tones for a while, wrote some instructions on a slip of paper, and then gave her a small bottle of children's aspirin to lower the fever before calling the next patient.

An hour later, the crowd dispersed; they would return later in the afternoon after lunch and a siesta, when the worst of the midday heat abated. Señor Vidal and I made our way across the village, passing the concrete-block school and grass soccer field, following a well-worn dirt path to his home, where his wife had prepared a meal of rice, chicken stew, and boiled green plantains. We climbed up the wooden ladder to his elevated hut and sat in the shade to talk and eat.

Señor Vidal was in charge of the medical care of 2,777 people spread among eleven village communities. Government initiatives to promote the health of rural communities involved establishing a network of sanitarios like him, stocked with basic medical supplies. Under this modern system, Señor Vidal, who had the medical training of a certified nurse's assistant, was charged with the unenviable task of serving as doctor, pharmacist, and dentist to a large population, while having neither the expertise nor the resources to be truly effective. He regularly collected blood smears to check for malaria—which was rampant—but often had limited to no access to medication to treat anyone who had it. With the introduction of processed sugars to the region, dental decay had become a major issue, and later that afternoon I observed him remove abscessed teeth from many patients who tried very hard to remain stoic without local anesthetics.

A short man with jet-black hair and glasses, Señor Vidal was a dedicated medical provider who clearly cared deeply about the health and well-being of people who lived in his district. He was diligent in his work and meticulous in his notes, but I could not help sensing a feeling of despair emanating from him with each case that he was able to diagnose but that a poorly stocked pharmacy made impossible to treat. The child he had seen earlier that day, he explained, needed antibiotics to treat his respiratory infection—he thought it was pneumonia. He gave the mother something for the fever because that was all he had to give her. He urged the mother to make the trek to Iquitos, the nearest city, to buy medicines there.

"Will she go?" I asked. He shook his head, looking in the distance, unable to meet my eyes, and said, "No, she will not. It is too far, the trip too costly, and the medicine expensive. It is out of their reach."

"What about the boy?" I prodded.

"If we are lucky, he will survive. But experience tells me he will not."

I SPENT MY FIRST FEW WEEKS seeing what I could learn from the communities closest to the lodge, the same one I'd stayed at over the summer, where I was glad to be reunited with Don Antonio. My return to the garden was met with his warm smile and embrace, and he greeted me with the nickname he'd given me. "Cashuka, I've missed you!" I presented him and Gilmer with some small gifts I'd brought from Atlanta—a nice pair of pruning shears, a bag of sweets, and photographs I'd taken of them.

As if I'd been gone only a week rather than months, we easily fell back into our familiar patterns of comradery. On days that I wasn't traveling to the local villages for interviews, I stayed in the garden with Don Antonio and Gilmer, cleaning the grounds or helping them prepare herbal medicines. I even got lessons in traditional building techniques when it was time to replace their hut's roof. They taught me how to weave the large palm leaves into tight, sturdy mats. As I wove the roof "tiles" together, they worked on top of the hut pulling off the old parts and replacing them with the new. The older palm mats went into a heap to fuel campfires for cooking. While it was hard work, I did it with joy. I loved learning firsthand how to live off the forest resources, and this better prepared me to understand the flow of life in the neighboring villages.

At one such village, called Llachapa—the closest ribereño village to the lodge, a short canoe ride away—I met Patricia. She was a bright and entrepreneurial eleven-year-old girl who came to the lodge often with her little brother to sell necklaces and bracelets that she and her mother made from local seeds and palm fibers to the tourists.

Some days, we sat together on a bench in the shade chatting about local life, with the camp's "pet," a capybara named Charlie, curled up in the dirt resting at our feet much like a dog would. It took me a while to get used to Charlie. The world's largest living rodent species—basically a monster-size rat—the capybara can weigh up to 150 pounds. They are a prized source of local meat, and I'm pretty sure I ate capybara soup in one of the villages. As for Charlie, he was as sweet as a mild-tempered dog and enjoyed human company.

Patricia often accompanied me on trips that I took to the local communities along the Napo River. She wasn't the only one. Many days I had a small army of burgeoning ethnobotanists who piled into the motorboat with me as I went to speak with community health providers and village elders. Patricia was their quiet leader; her interest in learning more about plants ignited the curiosity of other children. She had olive skin, black hair that fell straight to her shoulders in a simple bowl cut, and deep brown eyes that exuded an intelligence and curiosity that distinguished her from her peers. She taught me so much about useful plants of the area and showed me how to use dye plants like huito (*Genipa americana*, Rubiaceae) to color the chambira fibers (*Astrocaryum chambira*, Arecaceae) used for making necklaces, and how to clean and eat aguaje fruits (*Mauritia flexuosa*, Arecaceae). She also brought me to her grandmother's house, where I was honored to be offered masato—an alcoholic beverage made from fermented manioc, also known as cassava (*Manihot esculenta*, Euphorbiaceae)—for the first time.

Native to South America, cassava was domesticated in Brazil more than ten thousand years ago and was a staple food for pre-Columbian Americans. Portuguese colonists introduced it to Africa in the sixteenth century, and now Africa accounts for more than half of the world's production. Even though cassava comes from rainy, tropical areas, it has the added advantage of not needing excessive amounts of water, making it ideal for growing in areas at risk of drought. Today, cassava continues to serve as the leading source of carbohydrates in the diet of people living in the tropics.

Patricia explained that after boiling the tuber, her grandmother chewed it sometimes with a swish of sugarcane juice and then mixed it in a large barrel. I later learned that this step is scientifically important, as the amylase (an enzyme) in the saliva starts the process of breaking down the starches into sugars that in turn feed the natural yeast from the environment, kick-starting a fermentation process and the production of carbon dioxide and alcohol by the microbes. The result is a mildly alcoholic slurry of starchy manioc and spit.

I was offered a serving in a bowl made of a dry gourd that had been split in two. The texture was creamy, almost like buttermilk, and while I was honored to be offered a bowl, in truth it was hard to get over the gut reaction to its preparation method.

This wasn't the only culinary adventure I shared with Patricia. I spent a rainy afternoon on the steps of the lodge under the cover of the palm-thatched walkway chatting away with her and other local village kids. To pass the time, they were eating some snacks—but not the type of snacks you would envision kids eating. They had found a nest of ants that featured large abdomens, which they would pluck up between their fingers and squeeze, forcing a creamy semisolid liquid to emerge. They would suck out the liquid, toss the rest of the ant body, and pick up the next one.

Patricia gestured to me to join them in sharing their happy bounty and so I did, curious about this jungle snack. The taste wasn't bad— almost acidic in nature, with a citrus twist (likely due to the formic acid used as a poison defense by the ants). And so I, too, passed the time waiting out the rainstorm while munching on ants. It gave me the opportunity to learn more about the children's daily lives as they talked and laughed together.

Later that afternoon, though, I soon came to regret my decision to join in on the local fare. Don Antonio found me leaning over a wooden railing in the open hammock shelter of the lodge in a fit of painful projectile vomiting. The ant snack had not agreed with my stomach.

He told me to stay there and that he would be right back. I wasn't going anywhere in any case, as my body continued to contract in painful dry heaves. He returned with a cup of hot coca leaf tea (*Erythroxylum coca*, Erythroxylaceae) and a large handful of long grass, which he crushed and twisted into a big knot. After waiting for me to finish the tea, he led me to a hammock, instructed me to lie down, and plopped the grass knot on my chest. I inhaled its potent aroma—it was lemongrass (*Cymbopogon citratus*, Poaceae) from his garden—interesting, botanically speaking, because though it originated in Southeast Asia, this species had spread to various tropical regions across the globe due to its medicinal value.

He encouraged me to take slow, deep breaths and rest. Before he left, he shook his head and gave me a fatherly lecture: "Cashuka, you have the heart of an indígena, the mind of a brujita [little witch]," he said while gesturing to my heart and then my head. Laying his hand on my still-cramping stomach, he continued with a laugh, "But you have the stomach of a gringa. No more ants, okay?"

"Sí, sí," I agreed, before falling asleep in the hammock for the remainder of the afternoon.

In 1933, A YOUNG MAN born and raised in Boston named Richard Evans Schultes, the son of a plumber, entered Harvard University with dreams of becoming a doctor. But when he took a course called Plants and Human Affairs taught by the legendary botanist and director of the Harvard Botanical Museum, professor Oakes Ames, his interests shifted to the botanical realm. Under Ames's mentorship, he dedicated his senior thesis research to the study of ritual uses of the peyote cactus (*Lophophora williamsii*, Cactaceae)—rich in the psychedelic substance mescaline—by the Kiowa people of Oklahoma.

Schultes continued in his graduate studies with Ames and focused his PhD dissertation on investigating the identity of the fabled Aztec teonanácatl, meaning "flesh of the gods," in Mexico. Known today as "magic mushrooms"—rich in the hallucinogenic compound psilocybin—he identified them as members of the *Psilocybe* genus, situated in the Hymenogastraceae family. He also determined the identity of another fabled ritual ingredient locally known as ololiuhqui to be *Turbina corymbosa* of the morning glory, or Convolvulaceae family. The seeds had psychedelic properties later determined to be due to the compound ergine—an ergoline alkaloid similar in structure to that of LSD (lysergic acid diethylamide).

At the age of twenty-six in 1941, Schultes decided to travel farther south, embarking on a journey to document the practical, medicinal, and ritual uses of plants by Indigenous peoples of the Amazon rain forest. In all, he spent more than a decade in the Amazon in nearly continuous fieldwork, collecting more than twenty-four thousand species of plants—including some three hundred species that were new to science. He worked closely with Indigenous groups and trained under several healers, covering vast areas of the Amazon, including Colombia, Ecuador, Peru, Bolivia, and Brazil. Uncovering new insights into the science behind the relationships people have with plants, he published many accounts concerning the preparation of curare poison, the source of the paralytic agent tubocurarine, which had revolutionized surgery. He reported on the uses of ayahuasca, providing the first academic reports of the hallucinogenic brew and its ingredients such as *Banisteriopsis caapi* and *Psychotria viridis*. For his many contributions to this field, he is widely recognized as the "father of ethnobotany." I was here, doing what I was doing, because of Schultes's pioneering work.

I'd spent months between trips to the Amazon poring over his writings—scientific accounts of how wild plants were used by local people for food, medicine, hunting tools, rituals, and more. I marveled at the images in his photo-essay book, *Vine of the Soul*, which depicted the intricate uses of many plants by many Indigenous groups. There were black-and-white images of Siona Indians using blowguns to shoot monkeys to eat, Yukuna Kai-ya-ree dancers bedecked in ritual masks colored with the dyes of wild forest plants, and a portrait of a renowned Kamsá medicine man wearing loops upon loops of vegetal necklaces and holding a shakapa bundle in hand. Schultes made contributions to science that few people in this world could fathom, uniting his love for nature and medicine in his studies, to eventually bring those experiences back to Harvard, where he trained and inspired generations of scientists to come.

Nearly fifty years after his initial journey to Amazonas, I found myself skimming across placid, muddy rivers that he, too, had once traveled. But, in that half century, much had changed. The ongoing and rapidly increased Western presence and corporate resource extraction through the arteries of the Amazon had left its mark on the land, leveling dense forests rich in biodiversity through clear-cutting and scarring the earth with deep gaping holes as part of gold-mining efforts. But it wasn't just the land that suffered; it was the people who lived on that land, who depended upon it for shelter and sustenance, not to mention the intimate cultural and ritual relationship they had with it. Mestizo people, first stripped of their Indigenous heritage through subjugation and enslavement during the rubber boom, and the more recent introduction of government-run schools and the general encroachment of capitalist economics, now lived in clusters along the main riverbanks, no strangers to the forces of international commerce and development.

In Schultes's day, these people would have still relied upon medicine men and on the natural world around them for their subsistence. Now, it seemed to me that they were caught somewhere between the old ways and the new. Many in these communities had lost the knowledge I sought about traditional plant medicine, and yet they were not fully enjoying the real boons of what Western medicine could offer. Señor Vidal simply didn't have the resources to do his job effectively.

When Western medicine was first introduced to the region through government outreach initiatives, the medical clinics like his were well stocked with medicines, including anthelmintic drugs that can safely remove parasites and vitamin supplements to treat the vitamin A deficiency of local children. But because most communities didn't engage in the cash economy, they rarely had funds to purchase new medicines, and thus the local pharmacy boxes were rarely restocked once they were depleted. To do so would require a costly and arduous multiday trip downriver to Iquitos to sell precious livestock for funds to buy the needed medicines.

Western medicine and schools are not bad things in themselves. They're good, and so are jobs and houses that can better withstand flooding; but what is bad is the wholesale loss of indigenous knowledge, entire systems of thought and specific medical practices vanishing from the face of the earth forever. I knew there had to be a balance, some way to take what's best of both worlds to create a more robust and dynamic approach to health and well-being that honored the important contributions of the old and the new, nature and science, traditional medicine and modern medicine. Wouldn't we be so much stronger as a species if we sought that balance?

At the time, however, I was focused on collecting what I could learn about the old ways of reliance on wild plants to meet these people's basic needs. And to do that, I realized, I needed to go deeper into the forest, farther up the river. While I had learned a lot from Patricia and the mothers in the surrounding villages, I sought even greater knowledge of plants and their uses.

TRAVELING BY DUGOUT CANOE along with one of the staff members from the lodge, I set out to visit my first Indigenous community down the Sucusari tributary, a Maijuna village. Our paddles splashed quietly into the dark brown waters, creating a steady rhythm, until our movements fell into synchrony with the monkey calls and birdsong hailing us from the emerald tree branches that hung above.

Reaching the village, we pulled the canoe onto the dark, sandy shore. Excited children greeted us, having seen our approach from afar. As we walked into the village, I felt unsteady on my feet. I realized my prosthetic leg was shaky. Had a bolt come loose or had I unwittingly walked into something? When I turned my head to check if there was something behind me, I saw a cluster of children running back, giggling: they had been curiously reaching out to touch the prosthetic, making it wobble as I walked. I must've presented quite a sight to them—a gringa with short, spiky sun-bleached hair, wearing shorts cut above my artificial leg with fake plastic skin on display.

Once I sat on a bench in the center of the village, I took my leg off and invited the awestruck children to inspect it. I gestured to them to pass it around. No longer timid, they eagerly ran their fingers over it, lightly jockeying for position, looking inside the socket and even trying it on with their little knees poking in.

Introductions were made to some of the village elders, and I asked their permission to speak to local mothers about health. The Maijuna are also known by another, more derogatory name that emerged during the invasion of colonists in the region in the 1800s and early 1900s. During this time, colonists took to calling them the Orejón, or "big-eared," due to a cultural practice among the tribe's men of wearing disks in their earlobes, which are exchanged with progressively larger disks as boys transition into manhood. Historically, they adorned their bodies with the black tattoo dyes of the huito fruit and the waxy red colorant of achiote fruits. The elders of the community I visited no longer engaged in this practice, and like the ribereños, they wore Western clothing (faded T-shirts, flip-flops, soccer shorts), though they still carried on the tradition of creating bags, hammocks, cordage, and fishing nets from the fibers of the chambira palm.

What I learned from speaking with Maijuna mothers confirmed what

I'd gleaned from my discussions with Señor Vidal and the nearly seventy mothers I'd already interviewed in the ribereño communities closest to the river and the lodge. One affliction stood out: children suffered continuously from parasitic infections. Most households had six or seven children, though many of the younger mothers interviewed expressed interest in access to better family planning options, wanting only two or three. Access to healthcare was difficult. For some, it took an exhausting three to four hours to reach the medical clinic as the most common form of local travel was by dugout canoe. Review of the clinic records confirmed that people living closer to the clinic sought out care with much greater frequency than those farther away. That meant that for many children, care was seldom sought until an illness reached an acute state, leaving them chronically infected and weakened. Sometimes, as in the case of the little boy Señor Vidal had seen at the end of his long day, it was too late.

After my visit with the Maijuna, I heard of another village that was farther flung. Some of the local children who had accompanied me on my other trips knew of its location and were eager to take me there. The children seemed to suggest that the Yagua were a little different from other riverine communities. When we arrived I saw that the children, as well as young and middle-aged adults, were clothed in Western garb, but many of the Yagua elders still wore the traditional skirts and headbands crafted from fibers of the chambira palm. Some wore fish-scale necklaces made from the giant catfish that traverse the river, dyed a beautiful vibrant red color from achiote fruit pods, while others painted their faces red with the waxy achiote seeds.

I was thrilled. Not because I wanted to stare at the elders in traditional clothing, but rather because I knew they'd have the rich traditional knowledge concerning local plants that I was hoping to find. As introductions were made, a local man, stout and muscular, offered to translate, as he was bilingual in Yagua and Spanish.

In one home, I chatted with women preparing masato in the traditional way, while others sat together, expertly weaving fresh-cut palm leaves to create new roofing materials for their homes. While I interviewed them about home remedies for common childhood ailments (coughs, stomachaches, diarrhea), one woman's husband returned from his forest hunt. He was about five feet tall, dressed in the traditional fiber skirt, and barefoot, the thick soles of his feet having built up over years of forest hikes. In one hand, he held what looked like a tall pole that towered over him, but as he drew closer I realized that it was a pucuna—a blowgun! In his other hand, he clutched an adult spider monkey—a large black, furry New World monkey with a long prehensile tail—that dangled limply in his grasp.

I noticed some movement—a small tuft of black fur squirmed on the back of the larger monkey, before popping up its head, revealing a pair of curious eyes that explored the scene of activity at the hut.

The hunter gently removed the baby from its mother and handed it to his granddaughter, who was delighted to have it as a pet. During other household visits, I had observed that the youngest female children in any family often took care of small primates as pets, but I'd never been there to witness the actual handover from hunter to child. This tradition wasn't just for fun; it was an important way for the girls to learn caregiving skills, and, as they grew older, this role transitioned to care of their younger siblings, too. The pet, once grown, also assumed a new role—it transitioned into a family meal.

The stocky hunter then gave the adult monkey to his wife, who got to work cleaning it for dinner. I couldn't stop staring at him, and he noticed. Or, rather, it wasn't him—it was what he was holding. The man sensed my fascination. He held out his blowgun.

I'd grown up hunting with Daddy and my cousins in the forests and swamplands of south Florida, waiting out deer in the early mornings perched in a tree stand and working with dogs to find feral boar that were destroying local landscapes with their ground rooting. Raised with a respect for wildlife and for the sacredness of the hunt, I was always taught to take only what was needed—eat what you kill. In this sense, even though we came from worlds apart in culture, language, and environment, the Yagua hunter and I had a lot in common.

I graciously grasped the blowgun, feeling its weight and admiring its craftsmanship. I must've been as wide-eyed as the little kids who had earlier studied my prosthetic leg. This seven-foot-long handmade instrument, constructed totally of local plants, could accurately hit a target that was at the upper reaches of the forest canopy—up to two hundred feet above the hunter's position! As I turned it over in my hands, I realized it was even more beautifully constructed than I thought: it was actually two separate halves of skinny palm tree trunks, sealed together perfectly with a black plant resin. At the top of the blowgun, where the hunter placed his mouth, he had attached a piece of wood in the shape of an hourglass, with a hole drilled down the center. In addition to the blowgun, the hunter showed me the three other accessories critical to his art. There was a foot-long palm leaf, shaped into a sort of cylindrical basket, with foot-long slivers of palm material, straight and rigid—the darts.

Next, I investigated his small pouch, roughly the size of a pincushion and woven with a pattern that reminded me of my grandmother's crochet projects. Some of the fibers had been dyed red with achiote, and the inside was stuffed with fluffy white fibers of another palm, which had the color and consistency of cotton. Lastly, there was a small, hard gourd, just large enough to fit in the palm of my hand, which was filled with a dark black goopy resin. This was the curare—the powerful arrow poison.

I could scarcely believe that I was holding a container of it in my hand! I had read so much about this fascinating brew in Schultes's writing. Throughout the Amazon, curare is made by boiling certain plants together in water until it reaches the consistency of a tar-like paste. The most important ingredient is the curare vine (*Chondrodendron tomentosum*, Menispermaceae), which contains D-tubocurarine, which causes paralysis when it enters the bloodstream. By blocking a key receptor at the neuro-muscular junction, where nerve signals are transmitted between motor neurons and muscle fibers, the compound causes extreme muscle relax-ation, so much so that it can stop muscle movements in the diaphragm.

Curare dart poisons have been a source of fascination to Westerners since the earliest days of European exploration of South America. Sir Walter Raleigh wrote about his encounters with it in the late 1500s. In 1832, the famous Prussian polymath and explorer Alexander von Hum-boldt was the first to document how the poison was prepared from different Amazonian plants. Explorer and naturalist Charles Waterton demonstrated that animals can survive a curare exposure if given artifi-cial respiration until the poison wears off. But we didn't understand how exactly curare worked until 1935, when Harold King of Sir Henry Dale's laboratory determined the structure of the active compound (D-tubocurarine), working with a museum sample as his source. During his years of travel and research in the Amazon a decade or two later, Schultes helped to identify at least seventy species used to produce the drug, with some recipes calling for up to fifteen ingredients.

The Western fascination with curare extended well beyond its utility in hunting. Around the time when Harold King isolated its active com-pound, doctors began using it as a muscle relaxant, and its application in surgery was soon tested. Eventually, chemists devised a synthetic deriva-tive of D-tubocurarine for use in hospitals. Anesthesiologists now use a careful balance of muscle relaxants acting on neuromuscular receptors to relax the tissues for surgery while also applying anesthetics to ensure the patient is unconscious throughout the procedure.

Standing there with the hunter by my side, I was grateful. I lacked the words to adequately convey what this moment meant to me. I'd been put under anesthetics and muscle relaxants for more than twenty surgeries already—in some cases for periods of hours—as the surgeons amputated my leg, rebuilt my hip, lengthened my femur, sawed off bone spurs, and straightened my scoliosis-curved back. None of that would have been possible without the curious pot of plant tar. I realized in that moment that I was holding history in the palm of my hand. *This* was how far true medical knowledge could travel, across the world and down the centuries—and it all derived from the careful selection and preparation of a few plants growing in this very forest. I'd been born in a lucky era of medicine, one that benefited from the rituals and work of shamans who never knew how far their wisdom extended or how many lives had been improved because of their work.

As we began to say our goodbyes and pack up our gear to head back to the motorboat with enough light to return to the children's village and to the lodge, the local interpreter stopped me. I didn't think we had time to stay for dinner and make it back before dark. It wasn't that, though. The man told me that I was to take the blowgun as a gift.

"No, no," I protested, "I couldn't take something so beautiful and valuable."

The interpreter explained that the man had completed constructing a new one and that he would like me to have his old one. I asked if it would be okay if I gifted him some funds in return. After exchanging the gifts and a new round of thanks and goodbyes, I headed back to the boat, now the proud owner of an authentic Yagua blowgun.

LATER THAT WEEK, after a visit to a local school in a mestizo village, I joined the teacher outside as we watched the kids play soccer, some of them working in special moves to show off for us while we cheered them on. A few minutes into the game, two of the older boys came up the hill with a bucket of water pulled from the river. The thirsty children swarmed to it, reaching in a hand to slurp up some water before returning to the game. It was clear this was a common practice.

This surprised me.

While many mothers I'd spoken to assured me that they boiled water prior to consuming it and that they understood the importance of this practice, in reality it just wasn't practical to do this all the time. A recent public health campaign undertaken by the government to educate communities about the importance of boiling water had not been successful. Everyone drank raw river water.

This was another hurdle for Señor Vidal's attempts to provide adequate child healthcare. He simply didn't have enough backup; he was all alone. Although local children ate a balanced diet of wild and cultivated crops full of vitamins and nutrients, they suffered from poor health. His records meticulously noted the diagnosis of a variety of ailments, including anemia, malaria, diarrhea, and respiratory infections, which affected 54 percent of the children under the age of one that sought care at the clinic, and were the major causes of infant mortality.

The most prevalent affliction for the children was chronic infection from intestinal parasites, a sickness that affects children in developing countries across the globe. Helminths like roundworm, whipworm, pinworm, and hookworm are especially common in the tropics, where the problem is exacerbated by lack of sanitation systems or access to clean water. The worms cause iron-deficiency anemia, vitamin A deficiency, diarrhea and abdominal pains, cognitive deficits, and a weakened immune system, which is especially dangerous for children living in malaria-endemic regions.

Transmission of these parasitic infections can be through drinking contaminated water or through skin contact with contaminated soil.

Hookworm can enter the body by skin contact, migrate through the bloodstream to the lungs, pass through the trachea and swallowed, and then enter the intestines, where the parasites cause chronic intestinal blood loss as they feed on blood and tissues. They also cause intestinal inflammation, diarrhea, and distension of the child's abdomen—giving the child the appearance of a bowed-out belly. The pain associated with this disease is known locally as *cólico*. Because anthelmintic drugs like mebendazole and pyrantel pamoate either were locally unavailable or cost money that most people in these small communities lacked, the majority of cases went untreated. Options were limited, especially for young mothers caring for young children.

I was there as an observer, hoping to document and better understand the interface between people, their environment, and their health. I saw a lot of suffering, many children who were sick but didn't have to be. There had to be a better way.

One thing that stood out to me as I visited each village was the abundance of medicinal species I had learned about with Don Antonio. In particular, nearly every village featured a centrally located ojé tree.

What if this were a solution?

Don Antonio had explained the use of the ojé milk to me, and it was clearly a remedy long used in this region as a means of deworming children. I began asking the mothers I interviewed about it, and while they understood that children who suffered from cólico could take ojé to treat it, they were afraid. They knew that if given incorrectly, ojé could make their children very sick (Don Antonio had impressed upon me the importance of dosage when discussing nearly all the plants, but ojé especially). Out of fear of potential side effects, the traditional remedy remained unused, and the children remained chronically ill.

I understood the mothers' hesitation at using ojé as well as the challenge of reaching a medical clinic or hospital many hours and an expensive bill away. But this was a tragic situation. A solution to some of their ailments grew on a tree in the center of their villages, yet many lacked the knowledge to use it properly, while the easy-to-use Western pills hadn't been made widely available, although biomedicine had clearly supplanted traditional medicine. This was maddening: mothers desperate to keep their children healthy, surrounded by the resources that would allow them to achieve that end, and yet crippled by the lack of knowledge that would have been passed down from generation to generation had it not been outcast by Western influence.

I remembered a quote from a team of medical anthropologists that I'd read before coming to the Amazon: "Sickness is not just an isolated event, nor an unfortunate brush with nature. It is a form of communication— the language of the organs—through which nature, society, and culture speak simultaneously."

In the Napo River basin, I learned that illness is complex. And so is medicine. The very concept of what it means to be sick and what causes us to be sick is tied to and constructed by the sociocultural mores and spiritual landscape of a group. These questions must be understood before one can ever really address what it takes to become well again in this setting.

Traditional therapies rarely consist of administration of a drug (such as a plant medicine) by itself. Instead, they are accompanied by other, less material therapies, often involving the spiritual realm and aimed at healing the mind, body, and spirit of the patient. This is where biomedicine and traditional medicine reach a major chasm.

The time I spent with Don Antonio working in the garden and visiting his patients helped me to understand this paradigm of sickness and healing. I found a true appreciation for holistic medicine, and the thought of pursuing an education in the practice of Western allopathic medicine began to lose its appeal. Likewise, my interactions with local health officials, children, elders, and community members made a lasting impression on me. I was starting to see the bigger picture and thinking about where I wanted to fit into it.

One late afternoon in a local village, as the dying sunlight trickled through the canopy above, I found myself staring at the amazing buttress roots of an ojé, tall and narrow like curving walls. The smooth, thick trunk shot upward before branches began to grow confidently outward, blending into the leaves and limbs of its neighbors.

I didn't have the knowledge of Don Antonio, and I knew I never would. But he had opened a window for me, allowing me a view of the medicine, people, and the world that I hadn't known in college textbooks or back in the ER in high school. Standing beside this mighty tree, a sapling many years before I was born, I felt a mixture of gratitude and ambition, the sense of possibility.

I couldn't quite put my finger on it, but this sturdy tree, naturally bearing compounds of great importance to human survival, seemed, as its leaves shook in a welcome breeze, to crystallize something very special for me, a way to think about what I might be able to do with my life. I was born in a tiny town in Florida with birth defects that had shaped my life and passions, and yet I was lucky enough to have made it here to the Amazon, on the precipice of completing college and going on to med school—if I got accepted—and entering into the adult world, where I knew I wanted to make some kind of difference. I understood the powers of both traditional and Western medicine: my life embodied their efficacy. And I wanted to continue that journey. I wanted to keep working *right there*, at that intersection, where I could sense possibilities could keep growing, just like this massive ojé towering above me.

HOMECOMING WAS BITTERSWEET. I knew it would be a long time before I could return to the Amazon—the long path of medical school, residency, and fellowship loomed ahead. Don Antonio's son, Gilmer, gave me a parting gift of a huito tattoo—a serpent that wrapped around the calf of my left leg, ending at my toes. With time, memories of the way the jungle smelled in the midday heat—steamy and deeply herbal in the mix of mud and plants that I brushed against on my walks—began to dissipate. My huito tattoo eventually wore off, much as a henna tattoo does. I marveled over the photos I had taken on slide film, sharing them with friends as we looked at them over a screen projector—Patricia and me sitting on a bench, some children playing with my prosthetic. A month or two passed by with the rush of my final semester in college. Then, one day, something awaited me in the mail.

This was it. I ripped open the envelope to find my letter of acceptance to med school. I'd always envisioned this day as one of great celebration. I had worked toward this goal quite literally my entire life. All of the sacrifices made, the painful hours studying for biochemistry and physics, the late nights spent drilling facts into my brain in preparation for the MCAT—finally, I had made it!

Except, in that moment, I was astonished to find myself utterly unenthused about the news. I should've been dancing around the mailroom, fist pumping to the sky, calling my friends to get ready for a night out to celebrate!

I got on the phone to call my parents and broke the news. "Momma, Daddy—I was accepted to med school." They congratulated me, which only made what I was going to say next harder. "But I'm not going. I want to study plants and ethnobotany instead."

There was a brief silence. I wondered if there was a technical issue. They then replied in unison, "You want to become an ethno-*what?*"

CHAPTER 4

An Unexpected Houseguest

In an epiphany, I tasted how food weaves people together,
connects families through generations, is a life force of identity
and social structure.

SEAN SHERMAN, *THE SIOUX CHEF'S INDIGENOUS KITCHEN*, 2017

The first scientific conference I attended was the International Society of Ethnobiology meeting held at the University of Georgia in October of 2000. I'd graduated that May from Emory with a double major in biology and anthropology, and worked with one of my mentors, Dr. Michelle Lampl, over the summer doing some editing and research for a new textbook. Dr. Lampl trained as an MD-PhD at a time when few women attempted such a feat. She was a fierce advocate for her mentees and challenged her students to read primary scientific literature and to think critically as we assessed papers on one study after another. Her paradigm-shifting research on the phenomenon of bone growth in childhood was inspirational; she was a role model for me, demonstrating what a woman in science could accomplish with a combination of hard work, deep thought, and ingenuity. She encouraged me to think deeper about the meaning of health, and all of the many facets it encompasses.

As Dr. Lampl was familiar with my data set from Peru based on our long discussions about research, she encouraged me to apply to present my work at the conference. So I went about writing my first scientific abstract, entitled "Plants and Pills: Health Consequences of Western Medicine in the Peruvian Amazon," and submitted it for consideration.

When I received the response that I'd been selected to give an oral presentation, I was delighted—and absolutely terrified. I'd need to prepare slides and have them printed to slide film for loading on a carrousel that would project onto the large screen in the auditorium—not to mention speak like an expert for fifteen minutes, followed by five minutes of questions from a crowd of tens, if not a hundred, of real-life experts. I wasn't ready for this.

Plus, while I was now a college graduate, I had only a couple hundred dollars in the bank. Employed part-time doing freelance work, I couldn't afford to pay for a shared hotel room at the conference center. I could barely cover my share of rent. I could, however, afford gas, which was only a dollar a gallon at the time. I woke up early every day and made the hour-and-a-half drive from Atlanta to Athens in my baby-blue Jeep Wrangler, the worn soft top flapping noisily in the wind as I barreled down I-85.

I didn't know anyone at the conference; this was another solo mission for me. Up until that point, the only other ethnobotanists and ethnobiologists I'd encountered were in my mind, a picture painted from the books and articles I devoured, a sepia-toned bespectacled and balding Richard Evans Schultes chief among them. Now, people who shared my fascination with the human-nature interface gathered from across the globe—from graduate students to postdocs and professors whose work I'd read time and again.

Each day was packed with captivating presentations on topics ranging from the use of medicinal plants in the tropics to the assessment of their biological activities in the laboratory. I delighted in every lecture, taking careful notes. In addition to the talks and poster presentations, there were also workshops concerning intellectual property rights, traditional knowledge, and benefit-sharing agreements. The most prominent of these explored the Maya International Cooperative Biodiversity Group, a bioprospecting project established in 1998 by Drs. Brent and Eloise Berlin.

Funded by the National Institutes of Health (NIH), the project involved agreements between the University of Georgia, an NGO formed to represent the Maya people of Chiapas in Mexico, and a Welsh pharmaceutical company. The project aim had been to document traditional knowledge with the prior informed consent of the Maya people and examine the pharmaceutical potential of their medical traditions. Yet, not long after its inception, the project came under intense scrutiny and criticism from Indigenous activists and Mexican scientists: Was this *bioprospecting* or *biopiracy*? Bioprospecting is the search for natural resources from which commercially valuable materials can be obtained, whereas biopiracy exploits these resources for profit and without the authorization of or compensation to the Indigenous people themselves. The main point of contention was whether or not individual Maya could expect payment for sharing this knowledge that might lead to new pharmaceuticals. *Were they being exploited?*

The discussions were heated—some insisting it was ethical bioprospecting with appropriate informed consent, others staunchly asserting it was exploitation. It was amazing to see activists, scientists, and Indigenous representatives of the Maya people all at the table arguing for and against both sides. At this stage in my career, I knew only the basics of the issue and hadn't had any opportunity to read the actual agreements or dissect whether or how equitable benefit sharing was incorporated. I stayed quiet during these roundtable discussions, soaking up details like a sponge.

In 2001, a year after the conference, the NIH withdrew its funding, and the project shut down. This case drew international attention to the challenges of bioprospecting, and for me, it highlighted the importance of clear prior informed consent and equitable collaboration with Indigenous partners, communities, and local scientists. It wasn't until 2010 that the Convention on Biological Diversity's Nagoya Protocol on Access and Benefit-Sharing laid a foundation of international guidelines on these complex issues, entering into force in 2014.

Beyond soaking up as much science as possible, I was on a mission to find opportunities for graduate school and research experience. Lunch was served at large round tables, offering abundant chances to meet different scientists. I ate the lentil soup and grilled cheese sandwiches with new groups every day, intent on broadening my network and learning more about their respective research and educational programs. I met people from all over the world, including Dr. Andrea Pieroni, an extroverted Italian with thick black-rimmed glasses and dark walnut hair that always seemed to pop out in disarray. Completing his postdoctoral training at the University of London, Andrea focused on the study of how people identified, collected, and prepared wild vegetables as food, and what the perceived health benefits of these particular species held for the people who consumed them.

On the last day of the conference, Andrea and two other lunch buddies I'd met asked for a ride back to Atlanta with me. They needed to catch their flights and a bus, which were all departing from my home city. I was happy to oblige—as long as they didn't mind squeezing into the compact Jeep, plastic top on the verge of flying away, luggage on their laps.

After making drop-offs at the Greyhound bus station downtown and then the international airport, one passenger remained.

"Is this not your terminal?" I asked Andrea.

"Oh, it is. Or will be."

"What do you mean?"

"I'm going to the airport, too," he said in his Italian accent. "But not for three more days. Would you mind if I stayed with you?"

That was a surprise.

"Sure," I said. After all, he, too, was an ethnobotanist.

I could never have anticipated the importance of that one-word reply and all that it led to. It was clear from the beginning that there would be no romantic interest but that I'd found a big brother who shared a love for nature.

We drove back to the three-bedroom apartment on North Druid Hills Road that I shared with three others—my friend Hannah, her boyfriend, and a college buddy named Bryan. We had an old futon in the living room that I offered to Andrea for the duration of his stay. I opened up the windows in the living room to air out the space.

It was October, almost Halloween, and the fall air had turned crisp. To impress my Italian guest, I made a big batch of spaghetti with meat sauce. I used canned sauce and way too much ground beef, because the final result was thick and chunky, more like the consistency of a middle-school cafeteria sloppy joe sandwich than a delicate pasta dish. Although he ate the full plate, I could tell it was forced out of politeness.

The next day, I took him to my favorite place in town, the Atlanta Botanical Garden. I'd spent a lot of time at the gardens as a college student, repotting plants in the back room as a volunteer. My special spot in the garden was the Fuqua Conservatory, its large glass house reaching more than two stories high, home to a delightful array of tropical plant species, including cola nut trees (*Cola acuminata*, Malvaceae), betel nut trees (*Areca catechu*, Arecaceae), black pepper vines (*Piper nigrum*, Piperaceae), and palms of many shapes and sizes whose large leaves spanned across the pathways and reached up into the canopy. Sometimes, standing in the observatory, I could find a moment of quiet without the chatter of schoolchildren and other visitors. In those moments, I closed my eyes and breathed in the humid air, thick with the smells of forest plants and the soft chirping of frog calls, and I could feel myself back in the Amazon. I reveled in that sensation of wonder.

As a fellow plant lover, Andrea explored the tropical conservatory, hands fluttering over the leaves of plants as we walked past, admiring the patterns and variations in shape. Among the array of European species spread throughout the garden, he rattled off their scientific names, sharing details of how some were used in European herbalism.

As Andrea's visit drew to a close, we talked about my education and career trajectory, as well as my deep desire to learn more about ethnobotanical fieldwork. I'd accomplished a lot in the Amazon, but attending the large conference had been a revelation. There was so much more training I needed to do—specialized methods I hoped to learn both through graduate school coursework and hands-on experience in the field.

Because of my last-minute decision to pursue a PhD instead of an MD, I'd missed the deadline to apply to programs that year. I needed to take the GRE and start contacting potential mentors. The process was going to take time, and I was behind.

Andrea explained that he'd secured grant funds to lead a research team on a project to investigate the wild food-foraging behaviors of an ethnic Albanian minority group in southern Italy. The project started that spring.

"If you can get yourself there and pay for your food, I can put you up. That's how you say it? Like you've done for me. Plus," he added, "it's Italy! Maybe you'll learn how to cook!"

I was sold.

There was only one problem—the eternal problem, in life as much as in science—money. I had none. My part-time jobs didn't supply me with a cash reserve, and I didn't feel comfortable turning to my parents, especially because I now technically had a degree that ought to secure me a job. They were already helping me with my student debt.

I learned hard and fast that funding for field research is difficult, especially when you are neither an enrolled student nor an established professional. I looked for any and every possible source, no matter the award size; submitted essays, my research plan titled "Ethnobotany of Ethnic Albanians in Italy"; and letters of reference from my former professors and from Andrea. My persistence paid off, and to my ecstatic delight, I was awarded two grants—$1,800 from my sorority, the Kappa Alpha Theta Foundation, and $1,000 from the Foundation for Science and Disability. This would pay for my airfare, food, and incidentals. With housing provided by Andrea, I was going to Italy for three months!

ANDREA WAS WAITING for me as I exited baggage claim at Leonardo da Vinci–Fiumicino Airport in Rome. Dressed in jeans, a red T-shirt, and a light black jacket, he was carrying only a compact blue backpack over both shoulders. That was the extent of his luggage! Much like my trip to the Amazon, I overpacked, one gargantuan duffel. The next leg of our journey was to take a train from the central station in Rome to Rionero-Atella-Ripacandida, a six-hour ride with station changes, but we'd still manage to arrive at our destination of Ginestra before nightfall. We stocked up on some sandwiches for the journey and he insisted I try buffalo mozzarella—a particular local delicacy—for the first time.

After catching up on things ranging from life to the project's goals, we eventually fell into a companionable silence, enjoying the views of the rolling countryside slide by. There were endless fields of wheat, interspersed with clusters of olive groves and rows upon rows of grapevines.

The olive leaves shimmered green and silver in the bright March sunlight, and clusters of wildflowers had begun to emerge in a kaleidoscope
of colors—yellows, reds, purples—with the early warmth of the coming
spring.

In the distance, mountains arose, some ragged with sharp edges and
small villages perched precariously on top, while others appeared lush
and green, covered in vast forests of chestnut trees planted centuries ago.
My excitement for the awaiting adventure battled my exhaustion. The
beauty and the gentle constant rocking of the train eventually lulled me
to sleep.

I got my first glimpse of Ginestra from the smudged taxi window, as
we coasted along a small country road that curved back and forth, hug-
ging the rolling hillsides filled with thistles and cane reeds and bright
bursts of yellow blossoming on the cliff sides, where ginestra (*Spartium
junceum*, Fabaceae) bushes—the namesake of the village—flourished.
With a population of just under seven hundred, Ginestra was a small
hillside village, all the homes connected to one another, wall to wall.
Flower boxes dangled off windowsills and balconies, adding bursts of
color to the cream-colored plaster walls, and all the roofs were made
from those unmistakable red terra-cotta tiles. At the lower edge of the
village, there was an old church and cemetery—which the locals called
La Madonna—while the main piazza was the small village's heart,
where the bar-café and the cigarette-newspaper shop were located. In
the cigarette shop, there was a dial-up connection for internet access, and
once a week, we would log in to download and send emails, paying by
the minute.

After arriving in the piazza, I dragged my capybara-on-steroids-size
duffel across the cobblestone road, trying to keep pace with Andrea, who
took off to find Massimo, the local man who would provide us with the
keys to the apartment that he had arranged.

Pushing aside the curtain of beaded strings that served as an insect barrier, I walked into the bar. Putting my Italian phrase book to use, I drawled to the bartender, "Una Coca-Cola per favore," my southern accent making the words come out more slowly than intended. As he poured the drink, I took a moment to survey the small room. "Limone?" he asked, returning my attention to the counter. "Sì, grazie," I responded as he carved off a thin slice of bright yellow lemon to float atop the drink.

Every man—no women were whiling away time here—watched me, baldly curious about who I was and wondering what on earth I was doing in their village. Their eyes lingered on my prosthetic leg, the dark tan-colored plastic skin showing from my capris pants. After a moment of silence, the chatter returned with a fervor—none of which I could understand—and more heads poked their way into the barroom door to get a look.

Andrea called for me, and I polished off the drink to join him and Massimo by the apartment, right around the corner. The thick wood door to the building was nine feet tall with no door handle, just a single affixed knob that we had to push open using a large skeleton key twisted just so. Entering the building, Massimo led us up the staircase to the living quarters. It was musty and cold—the smooth marble held on to the chilly dampness of the last days of winter—but I was surprised by how large it was. It was apparent that it had not been lived in for some time—dust, cobwebs—but that only gave it more of an old-world charm for me.

There was a dining room, with a small cot in it, connected to a small kitchen not much larger than a walk-in closet. The living room, with cream-tiled floors and smooth plaster-coated walls, was beautiful but completely devoid of furniture. Massimo said he'd secure cots for the other two team members, Sabine and Harald, expected to join us next month, along with some other visiting scientists who would come for shorter stays. Down the hall was the master bedroom, furnished with an antique king-size bed, two beautiful oak armoires for clothes, and a small balcony overlooking the rolling hills, like a patchwork quilt made of fields, meadows, vineyards, and orchards. In the distance, the neighboring village of Ripacandida sat on a hill, with a deep green forest at its borders.

I assumed this would be Andrea's room. After all, he was paying for the housing with his grant funds. But he insisted that I take the room with the bed. He was content with the cot in the dining room.

The next morning, I awoke to the baritone calls of vegetable vendors—piselli, lattuga, carote, finocchio!—driving their small three-wheeled carts through the piazza and down the narrow, winding streets. Before we shopped for food, Andrea and I made our way to the piazza bar to order a cappuccino and a flaky cornetto, bursting with rich cream. We bought peas, lettuce, carrots, and fennel—the vegetables that called me out of my sleep—from one of the cart vendors and then stopped at the local butcher shop to buy chicken and at the bakery for durum-wheat bread that smelled of malt and caramel, with a hard crust I easily snapped off to sneak a savory taste. This shopping experience was so easygoing and logical, and yet wildly different from anything I was used to. Here, the vegetables were ripe and freshly harvested and reflected what was available in season, chickens slaughtered the day before, bread baked that morning. Later, we met the local shepherd, and he offered us the welcome gift of a sack of ricotta cheese made from the sheep he'd milked only hours before! Massimo gifted us wine and olive oil stored in dark glass bottles harvested in his family's garden and vineyard.

Everything was so fresh, local, and delicious.

THE ELDERLY LADIES in the village doted on Andrea—he'd been here before to establish the field site—and they welcomed us into their homes, smiling widely, wrinkles dancing, whenever they saw him at their door. On most days, we met with elders to discuss the traditional names and preparations of local foods. The men showed off their garden vegetables and wild-harvested greens while the women would prepare all sorts of traditional dishes for us to taste and discuss. As we sampled ornate plates of bitter greens, sweet pastries filled with cherry jam, dried fig cookies, homemade grappa, and dainty cups of espresso, I recorded the conversations (with their permission) by audio and video, while Andrea led discussions and took notes. On some days, I encountered a new occupational hazard I'd never anticipated: locals pushed so much home-brewed grappa and wine on us that we were forced to stumble back to the apartment tipsy, in need of a midday nap. Other times, especially when visiting multiple homes, we were offered so many espressos, we would depart jittery and wild-eyed. Whether it was coffee or alcohol, they took great offense if we didn't consume all of the beverages they prepared for us.

We were interested in documenting not just the Italian names of foods and ingredients but the Arbëreshë—or ethnic Albanian—names as well. Based on Andrea's pilot study research, he knew that the Arbëreshë had some unique practices when it came to the acquisition and preparation of plant-based foods. Our research goal was to investigate their use of wild plants as foods—and especially as "health foods," which we termed "folk-functional foods." We knew some wild plant ingredients were harvested and consumed for their general health-promoting benefits. These could represent an exciting starting point for research on the prevention and management of chronic inflammatory conditions common to the Western world, such as cardiovascular disease, diabetes, cancer, and more. Over the past forty years, scientific studies on diet and nutrition had revealed strong links between diet and "lifestyle" diseases associated with certain foods and more sedentary behaviors. Could some of these traditional foods be used to improve the Western diet?

The Arbëreshë are descendants of Albanians who fled the Ottoman invasion of the Balkans, migrating to southern Italy in several waves from the fourteenth to eighteenth centuries. This harked back to a time after Constantinople—the former capital of the Eastern Roman Empire— had fallen in 1453 to the Ottomans, who ruled from the city until the early 1900s. For six centuries, the Ottoman Empire was at the heart of engagement between the Eastern and Western worlds.

Today, villages of these Albanian descendants are scattered throughout southern Italy, especially in the regions of Apulia, Calabria, Campania, Molise, Sicily, and Basilicata—where we were based for our study. The Italian Parliament has officially recognized the Arbëreshë as a historic ethnic minority group, and their language, a variation of Tosk Albanian that does not exist in modern-day Albania, is listed as endangered in the UNESCO *Red Book of Endangered Languages*. Ginestra, known as Zhura in the Arbëreshë language, is one of three villages located in proximity to one another near the dormant volcano Monte Vulture, pronounced *vool-ter-ay*. The other two Arbëreshë communities included in our study were Barile (Barilli) and Maschito (Mashqiti).

At first, Andrea took the time to translate the conversations from Italian to English for me as he was conducting the interviews, but this made the discussions awkward and slow, as he was also the primary interviewer. After a week, he simply stopped—his way of pushing me to learn the language faster. It was sink-or-swim immersion learning. I trod water.

While we focused on documenting food and medicinal plants that the Arbëreshë used, our big question was regarding how knowledge was transferred over generations within ethnic minority populations following a migration event. The transfer, preservation, and even transformation of knowledge is a central theme in the field of ethnobiology. It's what shapes human experiences with resources found in the natural world. There is a huge difference between someone looking at a biodiverse landscape and seeing many different useful resources—whether they be ingredients for food, medicine, or art—and someone looking at the same landscape as a sea of green without direct use or connection to their lives. This relationship between humans and nature determines how we engage with the environment. If people perceive value in the landscape, they tend to have a greater desire to sustainably manage those resources, as well as to share this knowledge with others.

Just as I'd seen the influence of Western medicine on the decline of traditional health strategies in the Amazon, I saw the unsettling impact of industrialization on health and dietary practices here in Italy. Communities were losing the basic knowledge of wild food and medicinal plants once so prevalent. Younger generations, who no longer made a living through work in the fields but rather in automobile factories and shops in neighboring cities, rarely prepared these culinary treasures. Like grains of sand, these remnants of the culinary past were slipping through their fingers, along with the endangered Arbëreshë language. Andrea and I wanted to save what we could through recording as many details as possible. Preservation of this traditional knowledge was a major goal of this work.

And yet there was a key difference between what I saw in Italy and what I'd seen in the Amazon. Particularly in the Arbëreshë communities, there seemed to be a more nuanced understanding of how eating plants played a critical role in their lives, both short and long term. In this edible Italian landscape, the Arbëreshë held a strong preference for wild leafy greens that the Italian communities did not. For the Arbëreshë, wild greens known as liakra sat along the continuum of food and medicine—sometimes used as one or the other, and even both depending on the context and application.

We learned a lot about these folk-functional foods from Zia Lina and her husband, Zio Faluccio. Both in their late eighties, they carried on many of the Arbëreshë traditions lost in younger households. They still spoke Arbëreshë at home and ate wild-harvested vegetables regularly. Speaking with them was like looking through a window into the past.

Andrea and I sat at their small kitchen table, covered with a lemon-yellow and sky-blue checkered tablecloth. Zia Lina scurried back and forth, bringing over dish after dish of various greens she'd prepared for our visit. She was short, her hair arranged in a braided bun loosely affixed at her nape, and she wore a red shirt and a long black skirt, with a cheerful blue apron wrapped around her waist. Her toothy smile was contagious as she eagerly presented one culinary treasure after another. She cooked on a wood-burning stove. She had neatly organized shelves of wild herbs she'd collected, including a fragrant species of wild oregano (*Origanum heracleoticum*, Lamiaceae) known as Greek oregano that grows in rocky outcrops in the countryside. I'd never tasted such potent oregano before! It made the stuff I had in my shabby pantry back home seem like tasteless grass in comparison.

Zio Faluccio came up from the cellar carrying a box full of different home-canned goods, various plant bits stored in recycled glass jars under olive oil or vinegar, and tart jams made from wild fruits. He worked in the early mornings in their small vineyard, where he not only checked on his grapevines and olive trees but also tended his vegetable garden plot.

I marveled at the array of bottles he gingerly set on the table. These weren't ingredients one could find in a grocery store; each represented a flavor unique to this particular countryside. Some were intensely bitter, others tart and sweet. Zia Lina passed me a plate of wild asparagus (*Asparagus acutifolius*, Asparagaceae), blanched and then fried with eggs. Next came a plate of bitter greens—the young leaves of purple starthistle (*Centaurea calcitrapa*, Asteraceae), locally known as drizë—which had been boiled and then fried in olive oil with garlic and red chili peppers.

We nibbled on a raw salad of golden thistle (*Scolymus hispanicus*,

Asteraceae), locally known as kardunxheljë, dressed simply with olive oil they'd pressed from their own trees. Zio Faluccio added what looked like little onions to our plate. As soon as I bit into one, known as çëpuljin (translating to "little onion"), I realized that it lacked the sulfur taste typical of an *Allium* and instead elicited a potent bitter response on my tongue. It was the tuber of the tassel hyacinth (*Leopoldia comosa*, Asparagaceae) and later lab tests undertaken by Andrea and his collaborators at the University College London School of Pharmacy confirmed its potential for health-benefiting properties against aging-related diseases due to its potent antioxidant activities. For the elderly population in the village, this and other wild plants were common fare that had sustained them through famine and war but were also tied to belief systems concerning the plants' health benefits.

The concept of the Mediterranean diet has received much attention ever since the seminal seven countries study began in the late 1950s, evaluating differences in diet and lifestyle in the United States, Finland, the Netherlands, Italy, Greece, Japan, and the former Yugoslavia as they related to cardiovascular disease. This study resulted in the publication of more than five hundred peer-reviewed research articles and ten books on the science of diet, lifestyle, health, and healthy aging.

The Mediterranean diet in particular was associated with a 39 percent lower coronary mortality risk and a 29 percent lower cardiovascular mortality risk in middle-aged and elderly European men and women. Diet is a difficult subject to study, as each individual's preferences vary, but in general, it was observed that the Mediterranean diet included high intake of foods like legumes, breads, vegetables, fruits, fats rich in unsaturated fatty acids (such as olive oil), a moderate consumption of fish, and low intake of meat and dairy.

In the United States, the takeaways in the popular press on diet, as well as the ever-changing diet fads, have focused on the antioxidant-rich polyphenols of red wine and the healthy fats of olive oil. What we were learning in southern Italy, however, was that while the consumption of vegetables was important to health among the Arbëreshë, the type and characteristic of the vegetables actually mattered a lot!

Wild bitter greens were perceived to yield the greatest health benefit.

As THE FIRST MONTH came to an end, my command of the language had greatly improved. As it did, I was able to set out on my own more often to lead my own interviews focused on the Arbëreshë names of food plants while Andrea went plant collecting in the countryside. I set up my video camera with Hi8 tapes in the central piazza, next to the bar. This was an area that was highly trafficked by the villagers and where many of the elderly men whiled away their afternoons, chatting with their friends or playing colorful card games like scopa on small tables set out by the bar.

One by one, I was able to speak to many of them on camera, making careful effort to capture the pronunciation of various wild plants that Andrea collected in the countryside. Things were progressing, except for one point of contention in the community—Andrea's and my relationship. As was the case in many small, traditional Italian communities, there were clear roles that men and women were expected to play, and we'd unintentionally broken some big gender rules.

While unmarried men gathered to play cards, horseplay, and drink beer together in the piazza, women (married or single) did not, and they were also never seen out past dark. Unmarried women rode in the car of a man (other than a close relative) only if the couple was engaged to be married. Unmarried men and women simply didn't live together, and women did all of the cooking and cleaning no matter what.

So here I was, talking with many men in the piazza by myself during the day and returning at night to share an apartment with Andrea, and when Sabine—the tall and lithe, beautiful blond Swiss ethnobotany student—joined us, that raised even more eyebrows in the village. Now Andrea had two women at the apartment! Some speculated that this meant I was one of the unfortunate young women who were trafficked as prostitutes in the region. Even the two local police officers kept a wary eye of suspicion on me when they saw me out and about in the village on my own. Luckily, the elderly women I'd interviewed with Andrea were eventually successful in tamping down the rumors, scolding their husbands and sons for suggesting such a thing. They insisted that *their* Cassandra wasn't there for the village men; she only wanted to learn about plants!

I was glad for the women's defusing the tension since, during my second month, Andrea and Sabine left for a couple of weeks to present at a scientific conference. I was now on my own again, but as this was a small village, I rarely ended up alone. When I wasn't out plant collecting, conducting formal interviews, or transcribing data, the aunties (or zie) of the village took me under their wings and into their kitchens! Day by day, meal by meal, I began to learn the art of proper cooking, starting with simple garden or wild food ingredients, small amounts of meat, homemade pasta, and culinary herbs for seasoning to create classic southern Italian dishes, whether pasta con pepperoni crici-crici (pasta with crunchy peppers, black olives, spicy sausage, garlic, and olive oil) or zuppa di papnul e fazul (corn poppy [*Papaver rhoeas*, Papaveraceae] and fava bean [*Vicia faba*, Fabaceae] soup). All of this was a giant leap forward in my own cooking skills, which had previously ranged from the preparation of microwave dinners to the kind of canned red-sauce abomination I'd prepared for Andrea when he stayed with me in Atlanta.

But it wasn't just cooking I was learning; it was the entire approach to food of the zie, which was much more holistic than I'd realized. Zia Giulia taught me how to forage for wild chamomile (*Matricaria chamomilla*, Asteraceae) and mallow (*Malva sylvestris*, Malvaceae), and prepare neatly tied-up bundles of the herbs for drying and storage for use in the winter months as a tea for respiratory and gastrointestinal afflictions. Zia Giovannina taught me to stuff figs (*Ficus carica*, Moraceae) with almonds (*Prunus dulcis*, Rosaceae), and then dry them onto sticks carved from the giant reed (*Arundo donax*, Poaceae). She used these in decoctions, first boiling the almond-stuffed figs for twenty minutes in water, then cooling them to drink them as a treatment for respiratory ailments in the winter. Zia Lina taught me how to preserve cherry tomatoes in the cellar and dry those tasty crunchy peppers for use throughout the year. Zia Fiorina shared her love of language and music, and taught me the traditional Arbëreshë songs and dances that she and her husband sang on different cultural festival days.

Unsurprisingly, my education in Ginestra was very different from the one I received during my time with Don Antonio, where he was the primary repository for healing wisdom. Here, so many people, and mostly women, knew ways to use wild and cultivated plants for everyday health complaints. As I learned more about local customs related to food and health, though, I did see a major similarity to traditional medicine in the Amazon—and it had to do with spiritual health and healing. While Dr. Basso, a very kind older doctor with a broad smile and a gentle voice, kept an office in the village to see patients, the zie explained there were some things he was simply incapable of treating. I was puzzled. Like what? Zia Francesca took the lead in explaining malocchio (evil eye) to me.

Evil eye is a complex, popular illness that refers to the ability of the human eye to cause or project harm when it's directed at certain individuals or their belongings—a psychosocial disease linked to jealousy. Whenever I complimented someone on an item in their household—which I frequently did in my first month in the village, as this was a common practice of polite discourse in the southern United States, where I was raised—it created a problem. Oftentimes, the owner of the item (such as a teacup) insisted that I take it. No matter how much I politely refused, they insisted. Zia Francesca explained that this was because I was putting them at risk of malocchio. I admired something, expressed my admiration, but I hadn't canceled out the potential for the evil eye with the statement Dio ti benedica (God bless you). By giving the item to me, they were protecting themselves. Besides bringing bad luck to their household, adults afflicted by the evil eye come down with a strong headache concentrated in the forehead and behind the eyes.

I was deeply embarrassed by my ignorance. Had I inadvertently been casting a curse on these same women who had so warmly welcomed me into their homes and taught me so much about food, health, and life? My face turned red with shame as Zia Francesca explained how the evil eye worked. She reassured me that they knew I hadn't realized, and that it wasn't intentional on my part. Finally, I understood why I'd accumulated such an odd collection of knickknacks and teacups as forced gifts from them!

Certain men and women in the community, called quelli che aiutano, or "those who help," were well known for their healing abilities. Zia Francesca introduced me to her neighbor, a woman named Zia Carolina, who knew how to treat the evil eye and was willing to teach me.

Like me, Zia Carolina was disabled and walked with a pronounced limp, having been injured in a bus crash as a teenager and subjected to subsequent surgeries. She kept her graying hair cropped short and she dressed in the typical attire of the elder village women—a black dress in mourning for her husband who had passed on years prior. She explained that anyone who wishes to learn to treat the evil eye must go to church on Christmas Eve and repeat the prayer—an oral formula used in the ritual healing procedure. As I came to interview more healers, I learned there were actually many variations on the formula, and that each healer had one he or she preferred. One of them went like this: "In the name of the Father, the Son, and the Holy Spirit. Three have cursed me with their eyes, with the heart, and with the mind. Three are those who grow: the Father, the Son, and the Holy Spirit. On top of a mountain were a cow and a calf—the cow grazing, the calf growing. You grow yours and I grow mine." Another, shorter formula, got straight to the point: "The eye that has offended you, the Father, the mind, the face—remove the eye, evil eye, remove the evil eye from my life."

One thing remained consistent—the sequence and physical act of the ritual. Each therapeutic session began with three prayers common to the Catholic religion: Hail Mary, Our Father, and Glory Be to the Father. These would be chanted three times each while the healer rubbed her thumb in the repetitive motion of forming a cross on the patient's forehead. Next, the oral formula preferred by the healer would be repeated three times with the continued motion of the cross over the forehead. The healing ceremony could be performed anytime, day or night, and usually only one treatment was necessary, although some with persistent symptoms would seek out treatment for a period of three, six, or nine days.

The evil eye was just the tip of the iceberg. There were many spiritual ailments for which people sought treatment from specialist healers. The door into a whole new world of healing opened to me, and I was curious to find out more about these other folk illnesses, their treatments, and what role—if any—plants played in these rituals.

Master healers like Zia Elena, who lived in the nearby town of Maschito, focused on ailments with a perceived spiritual causation. Despite being in her midnineties, Zia Elena was spry and cheerful, and she lived on a diet of local durum wheat bread soaked in herbal teas, since she had lost many of her teeth. Her hazel eyes held a gentle warmth that always made me feel welcome whenever I came to visit. Some of the folk illnesses she treated were hair in the breast (mastitis), crossed nerves, wind illness, rainbow illness, dead-fire illness, mumps, water in the penis (paraphimosis), toothache, fallen fontanel, and more. The names weren't literal; instead they described certain features of the ailment's cause or therapy. Fuoco morto (dead-fire illness), which presented as a skin disease, involved the use of a small flame in the healing ritual. Like Don Antonio, Zia Elena had few apprentices and wanted to ensure that her knowledge was secured for the future.

Some of her healing rituals included veneration of plants, and, not surprisingly, this was of great interest to me. Cigli alla testa (migraine) both lays the blame for the headache at the elderberry tree (*Sambucus nigra*, Adoxaceae) and prays to the tree for help: "Good morning, cumpa' Savuche [godfather/close relative elderberry tree], I have the headache and I give it to you. I tell you the truth and I promise you, in the fire I will not place you."

There's a common belief that burning the wood of elderberry trees creates this ailment in humans, and, as a result, the trees are protected from harm. In some cases, people give offerings by tying ribbons to the tree, for example, or they perform the healing procedure under one. When the locals believe that natural resources—a tree, even a spring pool—have fearful magical powers, it sometimes leads to the preservation of that resource. On the other hand, it can also be detrimental if they use that resource, leaving it open to exploitation. One need only look to the example of the illegal trade of rhino horns for use in traditional Chinese medicine as an example of this.

Dermatological ailments were by far the most complex for healers to diagnose and treat. Take, for example, mal vjnt (wind illness). The disease presents with small round inflammations of the skin, and the Arbëreshë believe that the patient contracts it when walking near places where someone has been murdered in the past. The healer's job is to first determine which malevolent spirit the patient encountered. This is important, as the healing ceremony must use a weapon similar to that figured in the original crime—usually a knife, pistol, or small ax. The weapon is then used to bless a glass of red wine that is mixed with gunpowder and then painted onto each of the inflammations in the form of a cross using a braided bundle of donkey tail hair. The oral formula calls on Christ to take the bad wind and leave it under the walnut tree of good wind. Like elderberry, the walnut tree (*Juglans regia*, Juglandaceae) is considered spiritually powerful.

As with the wind illness, fuoco morto presents with symptoms of pronounced red eruptions of the skin, but these are filled with fluids. Its transmission is attributed to encounters with the spirit of someone murdered by fire, and like wind illness, the "murder weapon," fire in this case, is used in the healing ritual. Neither the healer nor the flame ever touches the patient, but the flame of an oil lamp or candle is waved in the sign of a cross over each area of the skin eruption as healing prayers are chanted. The treatment is performed every night for six to nine nights, and the patient is prohibited from entering the church during this period, as it could result in blessing the bad spirit, causing the spirit to never leave the patient and the illness to never resolve.

While the etiology, diagnosis, and treatment of these folk illnesses fall outside the paradigms of Western biomedicine, I came to appreciate the importance of several aspects of the process. For one, there was prolonged personal contact between the patient and the healer, and a strong belief by both parties in the diagnostic and treatment regimen. Whether a placebo effect, psychotherapy, or self-resolving illness (as many skin conditions can be), the patients often expressed deep gratitude for the treatment and improvement in their symptoms. Whatever the helpers were doing, it usually seemed to work.

One day I asked Zia Elena if she might be willing to turn her healing powers on me. On that day, her ninety-five-year-old cousin, Zia Sylvia, was visiting, seated in a wooden chair in Zia Elena's small living room. She had a nervous energy about her that reminded me of a little bird, curious and observant. I'd met her before at her home during an interview with Andrea, when she shared her knowledge of wild edible plants and served us two generous glasses of wine that had turned to vinegar.

So as not to startle them, I slowly lifted my pants leg and explained that my leg had been removed as a child, and that I suffered from skin irritation from the heat and friction of walking. Was there anything that she could recommend to heal it?

After I removed the prosthesis and peeled down the silicone liner, Zia Elena came closer and bent her head down low, crouching at my legs as she laid her hands on my stump. She smelled of chamomile. She began a series of prayers, her fingers lightly massaging the inflamed skin as she spoke. Her sense of caring for me and desire to heal almost thickened the air. She called on saints and the Holy Trinity to help me, repeating Catholic prayers in sequences of threes, all while gently caressing and massaging my stump. For a moment, I was transported back to Peru, where Don Antonio had called on his revered spirits of the forest to heal as he chanted his own ritual procedure.

At the end, I took a deep breath and exhaled, opened my eyes, and thanked her profusely. Zia Elena patted my head and smiled, saying anytime I needed her, she was happy to help.

Her cousin Zia Sylvia, who sat nearby, shook her head and sadly told me with tears in her eyes, "Poor child, when you marry, on your wedding night, your husband will find out! *He will find out!* You won't be able to hide it!" I laughed gently and told her not to worry. I would find a man who would understand, who would love me as I am. She continued to shake her head from side to side, her expression forlorn on my behalf.

She wasn't convinced. And, to be honest, I wasn't totally convinced either.

Hiding my leg—or rather the fact that I didn't have a real one—had become an obsession of mine when I hit my late-teen years. I didn't think that I was unattractive, and there were lots of things I liked about myself: my athleticism from the time spent outdoors racing quad bikes, climbing trees, and riding horseback; and my up-for-anything personality. But between my weakened right leg and the S-shaped curvature that remained in my spine even after the corrective scoliosis surgery, I was, well, lopsided. I used to joke with my girlfriends that I needed a one-sided butt implant to properly fill out my jeans. My thigh, calf, and gluteal muscles were overdeveloped on the left side to compensate. The only way my physical therapists and I were ever able to make my gait look somewhat even was if I walked with my right arm raised up and wrapped over the top of my head with my right hand over my left ear. Not really a viable option if I'm trying to look like a "normal girl."

ONE EVENING WHILE ANDREA and the other scientists were away, someone knocked on the door down on the first floor. It was already late, past 10:00 p.m., so this was unusual. I stepped out onto the balcony from the dining room to look down to see who it could be. It was a boy, maybe twelve or thirteen years old. I asked what he needed, but he just laughed and ran away. I went back to work, or tried to, but continued to be interrupted in the same way over the next hour.

I was getting tired and ready for bed, but knew I wouldn't be able to sleep if this little knock-and-hide game persisted. I leaned out of the balcony and shouted out, asking for his name, "Dai! Basta! Come ti chiami, ragazzo?" He ran off again. I got into bed.

Five minutes later, more knocking. My patience ran out—so I ran out of the house.

Once the big door closed behind me, the boy scurried out of sight. I walked around the corner to the bar and tried to describe him to the man, asking if anyone knew who his parents were. They assured me that someone would take care of it. "Grazie mille," I said, and returned to the apartment, only to realize that in my rush to catch the boy, I'd left the big skeleton key on the table upstairs.

Porca miseria! I stood in front of the big locked door, trying not to panic.

I didn't know if there was a spare key, I had no way to reach Andrea, and I didn't know where Massimo lived. I walked back to the bar. There was someone I recognized—Alfredo. I'd first met him when I was interviewing his grandmother, a village healer. Alfredo always had an easy laugh at the ready and seemed like an all-around nice guy.

Sitting at a small table, he was playing cards with a man in his late twenties—six feet tall and muscular with long straight brown hair that reached his mid-back. I didn't know his name, but he was hard to miss. I'd seen him a few times around the village, riding his cherry-red Ducati motorcycle or taking his black-and-tan Doberman pinscher for a walk. Rumor had it he'd taken that very same 916 model Ducati down to the nuts and bolts, and then put it all back together again for fun. I'd met his mother, Milagros, who was from Spain; when I first arrived, I was able to converse with her in Spanish, which was far better than my Italian.

I approached the table and explained my dilemma. Could they help?

"Sì, certo," Alfredo assured me, and then introduced me to Marco.

They followed me to the sturdy oak door. After Alfredo inspected the lock, Marco looked up the stone wall to the second-floor balcony.

"I'll be right back."

Marco started scaling the wall. There weren't many places to hold on to—the stone blocks formed a nearly smooth surface—yet he somehow raced up the wall, hopped over the balcony railing, and opened the door for us a few seconds later.

Alfredo laughed and patted Marco on the back. "You're like a monkey, man!"

Marco smiled at me, a little shyly, but I could tell he was excited to have performed this feat for me. I was excited, too.

MARCO RETURNED A FEW days later with his mom, Milagros. They were surprised to find the apartment full of scientists, as they'd come thinking I'd be alone for Easter. Regardless, they invited me to join them for lunch the next day with their family. Was this a date? I wasn't sure. Dating wasn't really done here. In villages like Ginestra, courtship was done al passeggio—on long walks in public view of all of the elders in the village.

The next day, I walked down a narrow cobblestone street to the lower portion of the village to their house, which was actually two large apartments belonging to twin brothers who each married a Spanish woman. Marco's dad was one of the twins, and he worked as one of the two policemen based in the village (his twin brother owned a local grocery shop). I cringed at what he thought of me: he was one of the policemen who'd heard the prostitute rumors when I first arrived.

Milagros and Marco greeted me warmly, and as they invited me in, I realized this wasn't just a little lunch for three. There were his grandparents, siblings and spouses, and all their children—and there was homemade lasagna, roasted lamb decorated with fresh sprigs of rosemary, mixed greens, salads, and loaves of sliced bread. This was a feast!

The squealing children darted between their mothers in attempts to snatch a bite of the sweet desserts in the kitchen. Marco's two sisters passed around bottles of Aglianico wine that the family had harvested, pressed, fermented, and bottled. The garnet-colored wine was full bodied, rich in earthy tannins, with a hint of musky berry flavors. The room seemed to spin with energy and rapid chatter, punctuated by exaggerated gestures, and I was seated in the heart of it, my chair next to Marco's. I was so nervous, at least a quarter of my meal ended up on the napkin in my lap, as I couldn't quite make the fork reach my mouth. With a raised eyebrow, Marco noticed my nervous ineptitude. I was mortified.

Later, we went on a chaperoned drive with his sister Rosanna and her husband to visit a beautiful forest located near the village of Ripacandida. The woods were covered with a light dusting of snow. I was still confused about the meaning of this day. Had they invited me because the family, like many of the elder zie, had just wanted me to feel welcome? Or was this the start of something more with this man whom I was instinctively drawn to?

Over the next weeks, the answer became clearer. Marco came to see me at the apartment in the afternoons and invited me out for walks. On these walks, we talked and laughed; but as delightfully surprisingly, we also easily shared stories of our pasts and our hopes for the future. He was an athlete in high school, and held the regional record for the three-thousand-meter steeplechase for several years. He'd run in the Venice and Rome marathons, and when time came to complete his mandatory military service, he was recruited to the military athletic team after completing basic training. These days, he worked at a delicatessen in the nearby town of Melfi. I visited him there once, amused to see the line of women vying for his attention as they placed orders for razor-thin slices of prosciutto, mortadella, and salame picante or containers of freshly made mozzarella or the decadent cream-filled burrata.

One night Marco borrowed his cousin's silver sports car and drove us to the neighboring village of Scalera, where we joined Alfredo and another friend of theirs, Pino, and their girlfriends, to check out a local festival. The streets were adorned with bright neon lights, transforming the sleepy town into a carnival, complete with rides, games, and the rich caramel aromas of roasted nuts served in paper cones. In other stalls, food vendors displayed large pale yellow gourd-shaped balls of caciocavallo curd cheese hung by red strings, where it was slowly melting onto thick slices of freshly baked bread over the flames of a grill. Local musicians sang Italian hits on the main stage. Young couples promenaded arm in arm while parents kept an eye on small children who played tag in the closed-off streets, melting cones of gelato dripping over their sticky hands. Teenage girls clustered together in groups, giggling at the shouts and antics of the groups of teen boys who tried to get their attention from across the street. A bold display of booming fireworks lit up the evening sky as the night drew to a close, and that's when Marco pulled me closer, dipping his head for our first kiss. Warmth rushed up my spine, and my stomach filled with butterflies.

As the weeks went on, Marco took me for rides on his Ducati around the curving hills of the countryside, my heart in my throat with fear I'd fall off the back of the motorcycle from the speed! On our first ride, I probably left bruises on his waist from how tightly I squeezed him.

One day, as we strolled through his family's vineyard out of sight from the village, I tugged on his arm and implored, "Tell me more about this place. What was it like growing up here?" Their vigna was a large plot of land full of grapevines, olive groves, mixed fruit orchards—bountiful trees of fig, almond, mulberry, cherry, and plum—and a vegetable garden.

He laughed, replying, "Come on, I'll show you my favorite spots." We meandered toward a hidden enclave where a small creek cascaded down a thirty-foot cliff in a waterfall, forming a shallow pool of water below. "I spent much of my childhood here, around this place, exploring the wild, looking for plants, and hiking with my dog." He gestured at wildflowers, shrubs, and trees, sharing many of their scientific names with me, surprising me with his knowledge of botany.

"Whenever I was in training for my next marathon, I ran through these fields and into the woods like a deer bursting through the tall yellow bushes of ginestra flowers." He continued with a chuckle, "Like a deer, I was also covered in ticks I had to remove each night!"

"How'd you learn about all of those wild plants?" I asked.

"My grandma and grandpa taught me a lot. We always picked wild chicory to bring home to eat. I also had a small book with color photos of the wildflowers that grow around here. Whenever I found something new, I'd bring it back home to try to find its name."

Like me, he was a nature lover who enjoyed botanizing—finding pleasure in learning more about the plants that captivated our attention. We'd both spent our childhoods scampering across woods and meadows, climbing trees, and examining the different creatures we encountered, whether flora or fauna—only some five thousand miles and an ocean separated us.

We continued up a dirt path, leading us out of the valley and through a shady forest of turkey oaks (*Quercus cerris*, Fagaceae) with terrestrial Italian orchids (*Orchis italica,* Orchidaceae) popping out of the lush green grass in bursts of vibrant shades of purple. At the end of the path, a large sloping hill spread across the horizon, covered by rows upon rows of neatly positioned grapevines. Along the vineyard's edge, huge bushes of dog rose (*Rosa canina*, Rosaceae) and elmleaf blackberry (*Rubus ulmifolius*, Rosaceae) flowered and tall stalks of the giant cane reed (*Arundo donax*, Poaceae) made a border between this field and the next, swaying with the gentle afternoon wind.

"Beautiful," I murmured. It was all I could manage.

When I turned to face him, his smile was broad with pride. "I'm glad you like it. This is my vineyard, I bought the land from my grandparents and I planted these vines. They'll be ready for harvest in the fall."

We stood together in silence, his arm across my lower back, basking in the light of the late afternoon. I leaned my head against his shoulder.

He raised one of his large hands to cup my cheek and looked into my eyes. His nails were trimmed short, and his hand felt rough, all that tilling in the vineyard and tinkering with his motorcycle and tractor in the garage.

He kissed me then, his lips soft and without urgency, as if we had all the time in the world to stand in each other's arms amid the grapevines.

But time, and its passage and rhythm through the seasons, gave me pause. We were quickly slipping from spring into summer, and toward my imminent return to my other life on the opposite side of the world. It was a reminder that I wouldn't be here with him when these gorgeous vines laden with fruit were ready for harvest.

My work with Andrea continued to progress with startling efficiency despite the occasional post-wine naps or the too-jacked-up-to-work caffeine highs. Once he returned from London, we organized everything we'd done and saw that, based on all the data collected on liakra, we had enough to begin the first drafts of our collaborative paper while still in the field. We'd also been gathering data on medicinal plants and were close to completing that survey, too. I'd logged more than a hundred hours of tapes from interviews with various healers throughout the region, and knew it would take months still to sort through the details of the intricate rituals of spiritual healing. That was work I'd have to complete back in Atlanta. At night, though, after our long days of research interviews and plant collections, I'd confide in Andrea about the blossoming romance with Marco. Not simply a research partner, Andrea had become my confidant, a treasured friend. He continued to share his expertise in not only field research methods, but also in the writing process as we worked on our research manuscripts.

I was eager to publish these results but worried about finding the right fit in a graduate program. "Any graduate program would be lucky to have you!" he'd reassure me. He suggested I try applying to London, where he was currently a postdoc, in addition to schools in the United States. All of this reminded me that I'd soon be returning home, that this little life I'd created for myself wasn't permanent, it was just another field expedition. I tried to push those thoughts out of my mind and simply enjoy what I had while I still had it.

One weekend near the end of my stay in Italy, I took a break from research work, and Marco stole me away on a trip to Maratea, known as the Pearl of the Tyrrhenian, located just south of the Amalfi coast. He packed a simple picnic lunch of ciabatta with fresh Italian deli meats and cheeses, a bottle of sparkling water, and some Aglianico wine from the family cantina for us to share.

We parked the car near a trailhead that led to a curving path along the coastline, and hiked along the mountain trails, stopping to look at the different plants we spotted in bloom. After passing through a deep green forest, we emerged atop a rocky cliffside that dropped off into the crystalline blue Tyrrhenian Sea.

"Come on," he said. "I know of a path down to the water."

We stripped down to our swimsuits, and Marco offered his arm to steady me as I took off my leg to leave behind on a boulder. We joined hands and jumped; I squealed with glee as we splashed down into the water. The sea was calm that day and gave us the opportunity to admire the views looking upward at the path we'd taken down to the shore. The mountains reached high out of the sea to kiss the aquamarine sky above, and the cove was quiet. We had it all to ourselves as we floated up and down with the gentle swells of the waves. It was our own secret, a wild hideaway from the world.

When it was time to get out, I sat at the water's edge as Marco jogged up the path to our boulder and grabbed my leg, stump liner, and a small towel to bring back to me. I patted the salty water off the skin of my stump and rolled the sleeve back on, allowing me to click the leg in place before we returned to our picnic spot. We happily dug into our picnic as we basked in the sun to dry off from the swim, our food spread across the smooth gray boulder.

He'd cut his once-waist-long hair to his shoulders, and it was slicked back, looking even darker with the salty seawater. I took a moment to watch him as he lay on his side, gazing into the distance, enjoying the view of the waves meeting the shoreline. He wore a red Speedo; his olive skin glistened in the sunlight. We were comfortable together whether in chatter or silence.

I'd come to realize that this man made me laugh and smile more than I had in a long time. And there was intense passion, a constant thread of sensory-filled tension that ran between us.

Despite my worries about my leg and some unfortunate experiences with boys I'd dated in the past, he put me at ease. I didn't feel self-conscious about my prosthetic or my surgical scars with him at all. It wasn't a big deal to him. *It was a nonissue to him*, never something he had to get over or accept. He saw me, the real me. Our attraction was mutual and fervent.

WHEN THE DAY of my departure finally arrived, I said my goodbyes to Andrea and the many zie who had welcomed me into their community. Andrea was going to stay on for another couple of weeks to wrap things up before returning to London. We made plans for finishing the remaining papers together via email in the months to come.

Marco drove me to the train station in Foggia. For once, the silence in the car wasn't simple and easy. It was loud—at least in my head. All the things I felt I wasn't saying but should banged and clanged. But what could I say, really? For all that I'd seen, learned, and experienced during my three months in this small town, meeting Marco had been the most totally unexpected and yet wonderful part of my fieldwork. But this *was* just fieldwork, another part of my education, my slow and steady march to graduate school and maybe someday to professorship. I'd proven Zia Sylvie wrong: I had found a man who didn't think my leg was a deal breaker. But, in another sense, she was right. I wasn't destined to be with this man. He had his job to return to, his fields to till and crops to harvest, and I had mine.

Marco stood on the platform with me and held me in a tight embrace as we waited for the last possible moment for me to board the train to Rome. As we kissed goodbye, I knew in my gut that this would be the last time I'd ever see him. This memory and so many others would slowly morph and merge into one stained-glass memory—maybe the swim in the Tyrrhenian, or the night he scaled the wall of my apartment—that I'd gaze back on periodically, triggered by the whiff of chicory or the top notes of a red wine, that curious spring in my twenties when I spent three months in Ginestra and met a man who made me feel so very alive.

As the train pulled away from the station, I pressed my cheek against the cool window, watching him until I could no longer make him out. I knew he was still there as the train gained speed, but I couldn't see anything through all my tears.

PART II

Infection

Rubus ulmifolius

CHAPTER 5

Wash and Fold

Yet the timeless in you is aware of life's timelessness,
And knows that yesterday is but today's memory and
tomorrow is today's dream.

KHALIL GIBRAN, *THE PROPHET*, 1923

My gut was wrong. I did see Marco again (he came to see me a few months later, arriving the day before 9/11 happened)—and again (we then went to Belize because of his visa issues)—and again (I eventually went back to Italy but had to leave because of *my* visa issues). In the end, it turned out we couldn't stay apart. But we also couldn't stay together: every time we saw each other, we ran into the time limits on our tourist visas, as if our two countries were the Montagues and the Capulets.

There was no starlit or beachside proposal, a fancy ring or bended knee—just one day in Rome, as another of our sojourns was set to end, Marco said, "We should get married." It wasn't exactly storybook, but in a style that has never failed us, we debated the merits and risks, shared our fears about never having lived with a partner before or knowing what marriage really entailed, and came to a decision: we loved each other enough to make this leap together. "Yes, let's do it," I said. "Let's get married." He lifted me into his arms as we kissed.

On the first day of spring in 2003, at the age of twenty-four—just two years after that trip to Ginestra—I stood in a thirteenth-century castle in Avigliano and pledged my sacred vows to Marco, a bouquet of miniature pink and white roses intertwined with hunter-green ivy in my hands. The civil ceremony was officiated by the local mayor, and in true southern Italian fashion, it was followed by eight hours of celebration, food, music, and dancing. In keeping with tradition, all the wedding guests gave us money, and by the end of the night we had more than €20,000, more than enough to pay the restaurant, the florist, musicians, and photographer, plus pay for a honeymoon and save the rest for our next steps, wherever we decided to live.

And where we decided to live, for the time being, was Arcadia.

Marco and I moved back to my hometown in Florida and into the back room of Daddy's ranch-style house. We had a nice little nest egg of cash from Marco's work savings and the gifts that we received—enough to buy a used car and have some in reserve for what came next. That was a little unclear. Marco possessed a keen knack for repairing engines and equipment of all sorts, and so he went to work for my dad repairing heavy equipment and helping out with land-clearing jobs. Daddy was happy, Marco was happy, and all was well—except for the prospects of my studies, career, and future. The previous year, as I readied graduate school applications, I'd taken a job teaching seventh-grade science at a local school. And to my great dismay, I hadn't matched with any graduate programs in the United States, and the program in London, which Andrea had encouraged me to apply to and where I had been accepted, would have put us into debt—or maybe even Chapter 11 bankruptcy.

I'd made attempts at applying to training fellowships from the NIH to cover my tuition and living expenses, but I didn't have any guidance through the grant-writing process and made several simple mistakes; the reviewers pounced on them. While they'd been impressed by my publication record at such an early training stage, they weren't fans of my research plan, which was exploratory rather than hypothesis driven. One reviewer wrote, "This might work out and she will find something. Or not. There is no hypothesis in the proposal. It is descriptive science and a fishing expedition." The program in London was a no-go.

One day, Sabine, my Swiss research collaborator from Ginestra, sent me an email about a new fully supported graduate program in ethnobotany based in Miami at Florida International University. A biology professor there, Dr. Bradley Bennett, along with several colleagues, had secured a training fellowship grant—a T32 award—from the National Institutes of Health to train a set of five doctoral students in tropical botany and ethnobotany. To me, it was like an omen . . . fate . . . destiny . . . surely a sign! There was one problem. Interviews and recruitment had happened in the early spring; it was June. Despite this tiny hiccup in the space-time continuum, I gave Professor Bennett a call, sent him my résumé by email, and scheduled a time to meet him in person, making the drive three hours south of Arcadia to Miami.

I was very nervous when I met Professor Bennett. He was descended in the academic training lineage from the legendary Richard Evans Schultes, completing his postdoc with Dr. Michael Balick, one of Schultes's former students. So many of my dreams rode on this meeting—that is, if there was a snowball's chance in Florida of their even considering me. A large bear of a man at well over six feet, Professor Bennett wore khaki shorts, a bright pastel floral Hawaiian shirt, and a pair of well-worn leather loafers. This was not typical professorial tweed. His brown hair was speckled with a few strands of gray, and dark suntanned laugh lines crinkled alongside his eyes. A native to south Florida, like me, with an easygoing demeanor, he welcomed me into his office, crowded with a

wall of books and various gourds and carvings one might find in a natural history museum. Photos of his family, featuring three young girls and
his wife, hung on the wall. An acoustic guitar was propped in the corner.
I settled into a chair next to a stack of scientific articles.

After exchanging pleasantries, he spoke about his ethnobotanical research with the Shuar people—also known as the Achuar, the infamous
so-called headhunters of Ecuador—with zest and enthusiasm. I was impressed by his passion and knowledge. In addition to directing the training grant, he also told me about the new center that had been formed to
build bridges across botany, natural products chemistry, and bioactivity
studies. The Center for Ethnobiology and Natural Products (CENaP)
was to be the home of some exciting work on local subtropical species in
the search for new drugs and validation of Indigenous medical traditions.

So now *all* my dreams were riding on this meeting.

Professor Bennett seemed to admire the work I'd already done, but he
did point out that I'd missed the deadline by not days or weeks, but
months. Still, he suggested that I complete the full set of paperwork and
gave a vague comment that, perhaps, the biology program might still be
able to consider it. I returned to Arcadia, assembled my documents, and
sent them off with a wish and a prayer. Maybe they'd consider me for the
following year.

I came home one afternoon in early August to find a letter from FIU
sitting on the kitchen countertop. I stared at it for a few moments, took a
deep breath, and then ripped it open. My eyes danced across the page
before the paper slipped out of my hands onto the kitchen floor. I had
been accepted into the PhD program as one of the five fully funded
scholars under the training grant. To accept the award, however, I
needed to start that fall. I needed to be in Miami in two and a half weeks
for the start of classes.

When I told Marco the news, he was ecstatic on my behalf, and I cried.

"Why are you crying?" he asked.

"Because you're so happy for me."

"Of course I am happy for you. But there's no time to cry. We need to pack!"

As usual, he was right.

¡Bienvenido a Miami!

MIAMI WAS SOMETHING OF a second home for me, my having frequently visited there throughout my childhood. Momma grew up in Miami Springs, and her brother and mom, my uncle Alan and Granny, both lived in Coconut Grove, a hip neighborhood located down by a sailing marina with a central shopping and movie theater plaza, bordered on its edges by rows of historic homes. My uncle was a successful chemist in his youth, but ultimately left the industry to pursue his dreams of owning a sailboat and making a living repairing and chartering boats. When he wasn't on a sailboat, he was managing the Bird Bath, a coin laundromat he owned in the heart of the Grove. Momma didn't think much of that; she always viewed it as a waste of his immense potential and talent. I saw the allure, though, even as a young child—and especially as an adult. There was joy in the freedom of sailing on the open ocean, the wind in your hair, the salt in your breath and coating your skin. There is nothing like sailing to simultaneously stoke and satisfy one's wanderlust. And if a laundromat is what keeps the wind in your sails, then maybe that's the price of freedom.

Miami was a marvel growing up. Some of my favorite memories as a child involved me picking avocados, mangoes, and coconuts from trees around my granny's apartment building. Coconuts were the most fun on the sailing trips we took with Uncle Alan. My sister Beth, cousin Melissa, and I would play along the boat docks in the Bahamas by taking turns bashing coconuts into the concrete like hard basketballs, competing to see who could break one open for us to drink the juice and eat the inner flesh.

The great thing about having my grandma and uncle both in town was that Marco and I had a place to crash while hunting for an affordable apartment near campus. While I was ecstatic over being selected as one of the funded trainees in the program, the reality was that the National Institutes of Health student stipend at the time was mediocre at best, totaling $19,000 per year, basically putting us under the poverty line. To compound matters, because of Marco's status as a spouse to a US citizen and temporary visa holder awaiting a green card, he couldn't be legally employed, and we didn't want to risk any under-the-table work that might compromise his pending application.

Never one to be idle, Marco spent his days riding his bicycle across town to Coconut Grove, where he would help my uncle refurbish his sailboat, sanding down the wood and applying fresh coats of varnish. And of course because we were living on the cheap, Marco also took the lead as the chef in our kitchen—pasta and his stunning homemade sauce didn't break the bank—and I continued my culinary education as his sous-chef.

The first few weeks of graduate school at Florida International University were a whirlwind: meeting my fellow lab mates, classmates, and faculty and jumping into coursework. I attended research seminars on topics ranging from environmental conservation to infectious disease threats. My old nemesis, staph bacteria (which had nearly killed me at the age of three), was the subject of intense study as resistant strains—MRSA, or methicillin-resistant *Staphylococcus aureus*—spread, and epidemiologists raced to trace its movements in hospitals and communities.

There were classes in medical botany, taxonomy, local flora, natural products chemistry, and analytical chemistry. I quickly realized what I had only allowed myself to be suspicious of previously: for someone who desperately wanted to become an ethnobotanist, I actually knew very little about the true science of plants—botany! While I had much more fieldwork experience than some of my fellow first-year classmates, whom I could regale with tales of Don Antonio and Zia Elena, I couldn't tell them the difference between the Lamiaceae and Verbenaceae plant families, the molecular structure of alkaloids, or how to measure the masses of compounds in complicated mixtures. So I got to work.

Learning plant taxonomy was like learning a new language. There were loads of new plant anatomical terms I'd never heard of before— stipule, corolla, trichome, style. Then there were terms used to describe the myriad ways that plant parts appeared or were organized. Leaves were arranged in a compound, opposite, whorled, or alternate fashion. Their shapes were deltoid, acerose, linear, flabellate, lyrate, falcate, ovate. The venation patterns had their own set of terminology, as did the shapes of the flower petals. The textures of various parts could be glabrous, hirsute, pubescent, viscid. The terms rolled around in my mouth clumsily, like talking with cotton balls in my cheeks.

Our small class spent days touching, smelling, and tracing the outlines of curved leaves, sticky stems, and odorous blooms in the classroom and grounds at Fairchild Tropical Botanic Garden—an enchanting eighty-three-acre spread filled with tropical and subtropical plants. I enjoyed learning about plants in such a lush place—especially considering the lessons that students in the colder north had to endure, learning plant identification through the examination of bark and leaf scars. Instead, we had a bounty of sticky, smelly, hairy plant parts to work our way through.

When we weren't in the gardens examining tropical plants, we spent time in courses exploring the native flora of south Florida. At first, the dense pine and saw palmetto scrublands, salty mangroves and swampy everglade terrains looked like a homogenous sea of green to me, very familiar and yet indistinct. But, with time, I came to see the landscape in totally different ways. Getting down on my knees in what the yogi in me calls botanist pose—butt up, knees bent, head down to the ground, hand lens (a small magnifying glass) affixed to the eye—I was rediscovering that same sense of joy I'd felt as a child when exploring the secret world of microbes in a drop of pond water or my own spit.

My professors taught me to use the totality of my senses in this exploration, whether I was examining a small herb, robust shrub, or large tree. Taking a bit of the leaf material between my fingers, I crushed and rubbed it back and forth beneath my nose, taking a deep breath to coat my nasal passage with its aroma—whether light and fragrant or deep and pungent. In my field notebook, I recorded the colors, shapes, and textures of the specimens, as well as noted the habit (the size and shape of the whole plant) and the habitat (where it grew and what other plants were nearby). I then sketched out the key features of each species I examined. I wasn't exactly Leonardo da Vinci with my depictions, but my hand grew steadier and I learned to trust my eyes, which were becoming more discerning.

As the terms became more recognizable in the classroom, so did the plants outside of it. The system of binomial nomenclature—first developed within Indigenous systems of knowing and organizing nature and later popularized by the Swedish botanist Carl Linnaeus—gave structure and order to the chaos of plant life that abounded. There was a comfort in the dawning understanding I had of their order—their relationships with one another, and how those relationships were signaled to me through those distinct sensory characteristics that they flaunted, subtly or boldly.

Through exploration of the gardens and the south Florida wildlands, I began the habit of greeting the plants by name, as if I were a medieval knight running into an old friend or Gollum from *Lord of the Rings*. "Oh, hello, my precious *Coccoloba uvifera* of the Polygonaceae family!" Commonly known as the sea grape, this buckwheat plant proudly displays its broad leaves and fat grapelike clusters of green immature fruits along the salty, windy beaches of the coastline. "Good afternoon, *Asclepias incarnata* of the Apocynaceae family," I would say, pausing in botanist pose while in a sunny opening in the swamplands, gazing at the delightfully mauve blooms of the swamp milkweed that attracts kaleidoscopes of butterflies and humans alike.

There were more nefarious species as well, like *Ficus aurea* of the Moraceae (fig and mulberry family). Commonly known as the strangler fig, it starts life as an epiphyte and then sends down roots and encircles a host tree, often resulting in its death, leaving a ghostly hollow where the host once stood. Perhaps none is more loathed by locals as the invasive species Brazilian peppertree (*Schinus terebinthifolia*) of the Anacardiaceae, or poison ivy family. Initially introduced to south Florida from South America as an ornamental due to its delightfully deep green leaves (pungent with the smell of black pepper when crushed) and bright red clusters of fruit, Brazilian peppertree has since escaped cultivation, taking over wetland habitats within the Everglades. *What is a weed but a successful plant?* I didn't hate the Brazilian peppertree. I loved it and all the other species I was meeting in my exploration of the plant world.

In addition to my time training in the field, learning about botanical names of plants, and exercising my newly earned knowledge of systematics, I was also trained in how to assess plants' chemical makeup through the field of phytochemistry. The FIU CENaP was composed of a series of laboratories outfitted for the processing and examination of plant tissues for their chemical makeup and pharmacological activities.

To find out why a plant does what it does, you first need to know what it's made of. Plant chemistry is incredibly complex—those aromas emitted from crushed leaf tissues or from nectar-rich flowers are the result of a mixture of hundreds and in some cases even thousands of unique molecules present in specific ratios. It's an intentional perfume blend crafted by the plant for the purposes of defense against predators or attraction of pollinators and seed dispersers. Like the microscopic world of protozoa and bacteria that had first fascinated me as a child, I found myself drawn to this new world of the unseen. I wanted the veil removed; I wanted to learn how to read the signatures of those plants I'd come to know in the wild. This was my chance.

"After you filter the liquid extract from the plant material, pour some into the round-bottom flask," Tim, one of the postdocs in the CENaP labs, explained to the small group of trainees. There were three of us women, all newly minted graduate students eager to learn how to transform raw, dried bits of plant material into an extract for chemical analysis.

Tim was tall and reed thin, a red goatee pointing downward from his angular chin. Beneath his white lab coat, he wore a Metallica T-shirt and faded jeans. He'd trained under the infamous cigar-chomping pharmacognosist and master of plant chemistry Professor Norman Farnsworth in Chicago and had come to Miami to continue his training under Professor Bennett. Now, Tim would also be sharing his knowledge of pharmacognosy (the scientific study of medicinal drugs obtained from plants or other natural sources) with us.

"Only fill it halfway full," he cautioned as I carefully poured the dark green liquid through the small neck of the fat round glass flask.

"What happens if I add more?" I asked, curious.

"Well, then it will *bump*. That precious liquid you just filtered will shoot up into the machine under pressure, and you'll have a green, sticky mess to clean up."

Duly noted. He proceeded to show us how to attach the flask to the rotary evaporator, sliding the glass neck into the fitting on the machine, with the round flask partially submerged in a container of hot water.

"Do the honors," Tim said, nodding to me.

I flipped on the switch and twisted the knob to a mid-range setting, sending the round flask into a constant spin in the water. The heated water bath gradually warmed the liquid, causing it to bubble like a witches' cauldron. When combined with the vacuum's efforts to lower the pressure of the closed system and the cooling coils affixed to the top of the machine, the alcohol in the extract began to evaporate from the flask and then condensate on the stack of icy coils, dropping the clear liquid into another glass flask affixed to the unit—leaving the plant compounds behind in the spinning flask. It was *so* cool!

While organic chemistry exams had always been incredibly challenging in college—the didactic class itself roundly considered a premed weed-out course—I always delighted in the experimental activities we undertook in the companion lab course. So being back among familiar concepts in a lab was a welcome balance to the ever-expanding vocabulary I had to develop as I dived deeper into our plant taxonomy coursework.

As we processed the ground-up bulk samples of plants collected around south Florida, I noted that extracts come in many colors—pale to dark green, red or purple, but most often a dark greenish brown. After the liquid extract was dried down, what remained was usually a thick tar-like goop. To remove any remaining wet matter, we were taught how to dissolve the goop in water, freeze it in the deep freezer at minus eighty degrees Celsius, and then attach it to the lyophilizer (a freeze-drying machine) for the last step in the process. If we were lucky, it was a fluffy powder or sugary crystalline mass, but sometimes it returned to its original tar-like state smeared around the inside of the flask. We spent hours every day using long bent spatulas to remove the material from the flasks.

Once removed, the remains were then weighed and stored in small glass vials in the freezer until needed for chemical analysis or use in laboratory assays to assess their bioactivity. I learned how to read analytical data from chemistry experiments, decoding the foreign language of signals—squiggly lines the lab machines spit out in distinct patterns, with each inverted V-shaped squiggle representing a different chemical in the mixture. Like the taxonomy coursework, lessons in mass spectrometry (MS) and nuclear magnetic resonance (NMR) brought with them new languages of symbols and terms.

While the analytical part of my brain enjoyed working on these puzzles, my greatest satisfaction came from our course in medical botany. Finally, I had a better understanding of the order of plants, their names and their relations to one another; it was like learning all the names for the constellations after so many years of just seeing a wild mass of beautiful stars. And now we were pairing that know-how with lessons from chemistry and human physiology. The distribution of plant compounds into specific plant families made a lot more sense now that I grasped the evolutionary relationships between those families.

The premedical college coursework I'd taken at Emory also proved especially useful as we dived deeper and deeper into not only how plants synthesized defense compounds for their own benefit but also how those compounds acted on specific receptors in the body, eliciting a pharmacological outcome.

Plant compounds affect pathways in the human brain in fascinating ways. Take the psychoactive stimulant effects of coffee. They're due to caffeine, a compound used by plants to deter insects. The same is true for nicotine, a compound of the alkaloid class that also protects plants from pesky insects. Other plant compounds act on targets in human nerves, muscles, the gut, bones, and more. My interest in these pharmacological activities was always driven by my intellectual curiosity, but sometimes these things became personal, as was the case with foxgloves.

I loved my uncle Alan. He helped me pay off my student loans and provided me and Marco a temporary place to stay while we apartment hunted in Miami. While he was only in his fifties, his health had been declining over the past year. As he explained it, "There's just something wrong with my heart."

Despite his active lifestyle, Uncle Alan developed type 2 diabetes as an adult, which later led to the development of congestive heart failure. I can clearly remember the first time I saw him exhibit the symptoms—I recognized them from my old days and nights volunteering in the ER. Though he was tall and lanky, his limbs appeared unusually puffy, swollen. I pressed my finger into the swollen tissue of his lower leg, and the indentation remained—this was a sign of pitting edema. I urged him to go to the doctor right away to determine what the cause was and to seek treatment.

Congestive heart failure is chronic and progressive, gradually decreasing the organ's pumping power. For roughly half of all diagnosed cases, the prognosis is an average life expectancy of five years. But for those at a more advanced stage of the disease at the time of the diagnosis, 90 percent die within one year. In historical medical texts, the condition was known as dropsy—a swelling of the soft tissues in the body with the accumulation of excess water. In today's medical texts, this swelling can be due to several underlying factors, but the most prominent one is heart failure. As the heart weakens, the ventricles can't move enough blood through the body, and the swelling worsens as the body fluids back up in the tissue.

There are a number of therapies that are used to treat edema, all depending on the root cause. When it comes to congestive heart failure, one of the major drugs is digoxin. Digoxin is a cardiac glycoside compound derived from plants in the *Digitalis* genus, commonly known as foxgloves. Historically, the digitalis leaves—especially the common or purple foxglove (*Digitalis purpurea*, Plantaginaceae)—were prepared as an ingredient in medicinal teas for the treatment of dropsy. William Withering, an eighteenth-century English botanist and physician, is credited with the discovery of digitalis and its uses for this purpose. However, there's debate over whether he may have in fact first obtained the recipe from a healer who used the herb to treat *her* dropsy patients. Yes, Withering refers to the healer as "an old woman from Shropshire." I'd say if the old woman was anything like the zie I met in Ginestra and Withering was anything like many men in history (and some in STEM currently), then it's fairly likely he did indeed come by his "discovery" via this woman whose name has been lost to history. What's certain is that Withering provided the first detailed medical accounts of patient cases in which he used digitalis formulations to manage dropsy, opening the door for integration into Western medicine.

Digoxin, an active compound from the plant, was isolated in 1930. It works by affecting potassium and sodium in the heart tissue to yield a stronger contraction of the muscle, which fosters a more robust and steadier heartbeat. Just like the plant, which is poisonous if consumed at high doses, there are also toxicity concerns for this active compound—there is a narrow "therapeutic index," as it's called, just like there is with ojé in the Amazon, between therapeutic and toxic doses.

For my uncle, the medical interventions to strengthen his heartbeat and reduce his edema were unsuccessful. I sat with him for days in the ICU, keeping him company, ensuring that he was as comfortable as possible. All the while, his lungs were filling with fluid, making it difficult for him to breathe. It was agonizing to watch him suffer. Draining the fluid through aspiration brought only temporary relief. His limbs were pain-fully taut from the fluid buildup, and his shoulders and arms ached. The day before he died, he asked me to massage them to offer some relief. I think he must have known the end was near, because when I told him I'd be back the next day, he insisted that I not come but rather go to my class at Fairchild Tropical Botanic Garden. It was one of my daylong taxon-omy classes.

We were out on one of the trails inspecting the living plant collection when I saw his best friend coming toward me with a security guard in one of the garden's golf carts. I immediately knew what that meant. He was gone. The funeral was held a few days later. Alan was cremated and we took his sailboat out to the bay with his closest friends and family to celebrate his life and spread the ashes.

Of his major possessions, besides the boat, there was the laundromat to deal with. My cousin, his daughter Melissa, inherited the business. She lived on the West Coast and with an active career of her own had no way of managing it from afar. Granny was too old to take care of it, and the only offers they'd had to purchase it were quite low. They figured if they were going to sell it cheap, the benefit should first be offered to fam-ily. Marco's green card had just come through, officially giving him permission to legally work in the United States.

Marco and I thought about the challenges and potential of the laun-dromat. We talked it out like we always do—somewhat methodically, over food, with honesty and good faith. The prospect of owning our own business was incredibly alluring, and could give us the greater financial security we desperately needed. Then again, what did we know about running a business? Quaintly named the Bird Bath Coin Laundromat, it was situated in the heart of Coconut Grove. There were more than forty washers, including some large-capacity ones, and two full walls of twenty gas-powered dryers. The walls and even some of the machines were

painted in bright tropical scenes of birds and flowers—good for the eclec-
tic vibe of the Grove. Besides the washing and drying machines, there
were also soda and snack vending machines. These could also provide
some extra income.

Of course, we had no money to come up with any down payment, but
after discussing the idea with Melissa, we came to a solution that worked
for all of us. She would sell the Bird Bath to us directly with no down
payment, and we would pay it off, month by month, using the cash we
earned operating it. This allowed us to forgo the necessity of a bank loan
and keep the business in the family.

Marco took on the business operations, making it his full-time job.
Besides managing the employees, he continually repaired the older
machines—which, obviously, called upon mechanic skills he already
possessed in abundance. Granny helped us get the books set up, and I
took over the duties of managing payroll, paying the massive utility bills
for gas, electric, and water, and handling the taxes. When I wasn't working late in the lab brewing up plant extracts or studying, I joined Marco
in the evenings to fold customers' shirts, jeans, and underwear, placing
them neatly into tightly wrapped plastic bags for pickup.

With hard work, the laundromat brought us a measure of financial
security. But as with plant taxonomy, the learning curve was steep. Employees stole from the cash register, and we had to learn how to manage,
fire, and hire new people. Strangers robbed us, too. Despite our monthly
payments to a security system with motion sensors, and locks for the door
and windows, one of our first robberies somehow evaded detection and
resulted in massive damage to our equipment. We were devastated and
angry. Though robberies in the city were common, each time it happened I felt personally violated and emotionally battered. It made me
second-guess our decision to take on the business, and the damage cut
into our already razor-thin profit margins.

The video surveillance system showed the thieves wielding bolt cutters and a crowbar to break through a large ventilation fan system in the back of the building and then using the same crowbar to pry open our large change machine. We watched the black-and-white video in a state of disbelief. How could someone do this? It's not like our business had the appearance of wealth. Had the culprits been inside as customers in the days before, watching Marco at work as he tended the cash register or checked the cash machine? Were we throwing ourselves into unnecessary danger with this business?

Marco usually emptied the cash from the coin change machine each night (stacks of fives, tens, and twenties) but had forgotten the night of the robbery. Thankfully, the thieves didn't realize that the bills were in the machine. Instead, they stole a wash-and-fold customer's pillowcases, filled them with quarters, and sneaked out of the building with a bag of coins like a couple of evil Santas. The pillowcases happened to have been very expensive, and the wash-and-fold customer was incensed about their loss, so in addition to the cost of repairing the coin machine and replacing the missing quarters, we also had to buy a new fancy Egyptian cotton, high-thread-count sheet set.

COCONUT GROVE IS AN ARTSY part of Miami, full of brightly colored buildings, floral-scented tropical air, and saucy Cuban music that spills out of the shops and restaurants onto the street, beckoning you to add a little sway to your walk. One of my favorite spots in the Grove is the Kampong, a lush tropical garden that is part of the National Tropical Botanical Garden network of five gardens, with the others based in Hawaii.

The name Kampong comes from the Malay or Javanese word for a village. The Kampong is full of plants collected by David Fairchild, one of my botanical explorer heroes, on the extensive global expeditions he undertook in the late 1800s and early 1900s. Fairchild was responsible for introducing more than two hundred thousand exotic plants and crop varieties to the United States. Built on Fairchild's former estate, the Kampong is a living genetic repository of plant diversity and contains a veritable cornucopia of strange and alluring tropical fruits. There are more than fifty varieties of mango (*Mangifera indica*, Anacardiaceae) that Fairchild collected, and the garden also features jackfruit trees (*Artocarpus heterophyllus*, Moraceae) that yield the largest edible fruit in the world at almost seventy pounds per fruit, roughly the weight of a ten-year-old boy!

Every year in Coconut Grove, the community celebrates this botanical bounty by paying a special tribute to mangoes—the myriad flavors, colors, and shapes of the six hundred varieties grown at the Fairchild Farm. There is also an annual King Mango Strut Parade, a delightful event rich in satire where locals dress up in costume, construct elaborate floats, and create signs stapled to pickets, which are held high in the air as they strut down the parade route through the heart of the Grove. There is upbeat music and wild dancing in the street accompanied by plenty of laughter and a hodgepodge of parodies on the news topics of the year, whether related to politics, health, celebrity scandals, or pop culture.

One year, Marco and I enjoyed the parade as part of the audience. Coming up behind a green palm-leaf-covered pickup truck with SAVE THE EARTH signs was an unusual group dedicated to bacteria. The cranberry-colored Volkswagen Beetle at the heart of the group was covered in white posters with bold-print lettering that read DON'T *STAPH* ON ME and STAPH MEETING. A large decorative banner was stretched across the roof of the car on poles, the words stamped in bold black and reds: SUPERBUG MRSA MILITIA!! Bedecked in full-body white hazmat suits and medical face masks, the strutters shook milk cartons that rattled from the clink of coins, bold Rx prescription labels scrawled on the jugs. Others wore white plastic trash bags over their T-shirts, holding picket signs and chanting, "Today we infect!" and running up to members of the crowd of onlookers, shouting, "We're coming for you next!"

I'd never seen anything like this. How had staph infections become the subject of satire for this community parade? *MRSA*. This once little-known acronym used by microbiologists was becoming part of the every-day vocabulary. That's when I knew we'd reached a tipping point. I'd been reading about MRSA, and it had come up in one of my classes, so while I knew it was a major medical problem, its appearance in the pa-rade was intriguing. I really shouldn't have been surprised. A report from the CDC noted that based on disease surveillance, there were an estimated 108,345 cases of invasive MRSA infections and 19,479 deaths in the United States in 2006. To put this in perspective, that same year there were an estimated 14,627 deaths from AIDS. Today, while the CDC estimates that roughly 1.2 million people live with HIV in the United States, advancements in treatments and public health measures have lowered the number of deaths to 13,000–15,000 a year. MRSA kills more people than HIV in the United States, and has continued to do so since 2005. Most alarmingly, these invasive MRSA infections weren't just happening to people who were sick or under hospital care for other conditions; they were also happening to young and healthy people. This became known as community-associated MRSA, or CA-MRSA.

Reports showed that student athletes in peak health were being in-fected: the frequent skin-to-skin contact and small abrasions acquired during football tackles, wrestling matches, or rugby scrums put them at highest risk. But even athletes in noncontact sports, ranging from base-ball to soccer, were being infected and falling gravely ill. And then there were infections tied to the gym and contaminated workout equipment. In 2004, a healthy eighteen-year-old Royal Marine scratched his legs on a gorse plant (a wild thorny shrub in the pea family) in England during an outdoor training exercise and died three days later. In 2007, a seventeen-year-old Virginia high school senior succumbed to his MRSA infection after a week, prompting the closure and deep cleaning of twenty-two schools in the region. Hospital outbreaks in pediatric units led to the deaths of babies and newborns. People living with HIV/AIDS were also under attack by CA-MRSA, both in the community and in medical care facilities.

Symptoms often began with a small bump or sore area in the skin that became red, warm, swollen, and painful to the touch. There would be pus and a fever. Sometimes, these bumps were mistaken for spider bites due to the small hole that could appear in the center of the swollen area, delaying a trip to the doctor. Once someone became ill and did go to the hospital, they were often subjected to aggressive and necessary procedures like debridement (the surgical scraping and removal of infected flesh) and long courses of IV antibiotics that battled the infection, but came with their own consequences of destroying the healthy gut microflora and causing painful bouts of diarrhea and vaginal yeast infections. In cases where the tissue damage was extensive, skin grafts were required to cover up large parts of the body where supple, healthy skin and soft tissues had decayed and rotted under attack from this microbial foe. Worse yet were the cases wherein the skin infection traveled deeper into the body, metastasizing by circulating with the blood or through forming a stronghold in the heart, spinal muscles, and bones.

For some of the most unfortunate cases, an infected scratch could lead to necrotizing pneumonia, which ravages the lung tissues, causing gangrene, followed by death in 61 percent of cases. These were the invasive infections that could take an otherwise healthy teen to death's door in a matter of days, and such cases were appearing in the news with greater regularity, stoking fears.

Attempts to develop a vaccine against staph infections were underway, yet more than a decade later, these have all continued to fail. Numerous clinical trials have shown that efforts to induce humoral immunity with staph vaccines were not sufficient to protect against infection. It is a wily organism, with a multitude of tricks for escaping immune notice, even hiding inside our own human cells!

With each headline covering the growing numbers of outbreaks and deaths due to MRSA, it was becoming clearer to me that new therapies were urgently needed—ones that could not only destroy these microbial foes, but also dampen their overwhelming toxicity and ability to overcome even the healthiest of immune systems.

I knew from my work with healers that many of the ailments they treated involved skin. I wondered, Could any of these skin ailments have been caused by staph? And if so, how did their remedies work? I knew they used plants in many of the rituals, but there were many other treatments for skin conditions that I'd not yet documented. From my graduate coursework in medical botany and survey of the ethnobotanical literature, I knew that the topical application of plants for various dermatological conditions was not limited to Peru and Italy. People all over the world did this, making use of the botanical resources available in their backyards and in the surrounding countryside and forests to treat their scrapes, boils, burns, and rashes. The pieces to the puzzle of what I should focus on in my doctoral research began to take on a more concrete form in my mind: medicinal plants, traditional healing of infectious and inflammatory skin diseases, and MRSA.

In 2005, at the age of twenty-six, in my second year of graduate school and marriage, I was on the phone with the orthopedic office—but this time, instead of leg problems, I had arm trouble. I'd slipped on a wet spot in the tile floor at home that morning with my fake leg, which flew out from under me like a rocket. I caught myself on the way down, my left hand taking the brunt of the fall. My wrist had been aching ever since.

"Okay, we can get you in to see the doctor this afternoon," the nurse told me. "You'll need an X-ray. Any chance that you're pregnant?"

Even though I was still in graduate school, Marco and I had achieved a semblance of financial stability with his running the laundromat, and we wanted to start a family. Both of my best girlfriends in Arcadia, Mandy and Jayme, had four-year-olds already, and I knew that with my constellation of birth defects and our desire to have a large family, we would need to start soon. We hadn't been on any sort of mission to get pregnant, but there was certainly a chance. I grabbed some home pregnancy tests at the corner store and brought them home. I bought a three-pack, just in case I needed it for the coming months. As I waited for the results, I called Mandy. "I think I may have broken my arm," I lamented. And as we chatted, I glanced over to see those two faint pink lines become darker and darker. I took in a deep breath. "And it looks like I'm pregnant!"

I jumped off the phone with her and called Marco.

We were going to have a baby! We had some serious planning to do.

We'd already given a lot of thought to how it might work before the pregnancy became a reality. From a timing perspective, it all actually worked out quite well within the system I had to operate under. The baby was due in August, and I could spend most of the fall and winter with him at home while doing remote work in planning logistics for my dissertation field research, securing permits and approvals from the institutional review board for the interview component of the project. Then the baby and I would travel to Italy, where we would stay with my in-laws while I undertook field research on wild plants used in traditional medicine to treat skin infections in the Monte Vulture area, using Ginestra as a home base.

Milagros had already offered to help me with childcare when the time came. Marco and I had long discussions on the best path for this aspect of my training, and we'd both agreed that while being apart would be awful (he would need to stay back to run the business), being with his family would make things much easier for me and he'd worry less about both the baby and me. We squirreled away as much extra cash as we could as the months marched by. If we stayed on track, we'd be able to afford day care when the time came for me to return to the lab after my fieldwork.

Months after recovering from the broken arm and carving off the cast, I was back to normal—albeit my new normal was a novel walking style to accommodate my growing belly. I no longer limped through the research building hallways; now it was more like a limping waddle, resembling an ungainly duck.

I was wrapping up my second year of coursework and basic training in lab and field techniques. Plant identification had become strangely easier as my sense of smell heightened—a surprise bonus of the pregnancy— which enabled me to sniff my way through plant samples strewn across tables for our taxonomy exams. While I couldn't identify species on smell alone, it certainly did help give me some clues as I carefully examined the structures of leaves and floral parts in the effort to group them in one of the four hundred or more flowering plant families that exist on earth. Exams were usually limited to plant families from the tropics and subtropics, but the number of possible options was huge to know by memory.

Following the completion of coursework, I had doctoral qualifying exams to prepare for. These are comprehensive exams that serve as the barrier between simply being a graduate student and becoming a doctoral candidate. The exam was prepared by my dissertation committee, composed of six members including an infectious disease neonatologist, a statistician, an analytical chemist, and a molecular biologist, and was chaired by Professor Bennett, my adviser and the ethnobotanist. In retrospect, maybe it hadn't been so wise to pick committee members from so many fields, because I had to demonstrate my competency across all of these areas both in long written responses that took days to complete and in an oral defense during which they volleyed one question after another at me. It was a hell of a week, and I really missed the stimulant buzz that coffee would have given me, prohibited by my pregnant state.

Successfully passing these exams and getting my research proposal approved would mark a major turning point in my graduate training. I'd be done with all of my classes except for one advanced course in plant taxonomy that I needed to complete, and could focus the rest of my efforts on preparing for my upcoming field research expedition in southern Italy the following spring.

WHILE I NORMALLY LOVED those hot Miami summers, at the height of pregnancy, I was utterly miserable. As the end of the pregnancy approached, Milagros came over from Italy to help us get ready for the baby. She pampered Marco and me with home-cooked Italian fare; she scrubbed our compact two-bedroom apartment from top to bottom, and helped us set up the last details for the nursery. We were ready and now it was just a matter of counting down the days until the scheduled C-section. Due to my hip dysplasia and surgical history on the bones of my pelvis, it would be physically impossible for me to experience a natural birth. If I were to attempt it, both the baby and I would likely die. The C-section would be necessary for us both.

A few days before I was due, news of a hurricane barreling toward the Miami–Fort Lauderdale area pervaded TV channels and radio waves. It was called Katrina. There were flood warnings, and as the Bird Bath was so close to the marina and water, we scrambled to come up with a flood plan to protect the equipment. We stocked up on drinks and canned goods, filled the tub with water, got propane for our camp stove, filled up the tank of our car, and closed the laundromat. We waited out the storm in our apartment, which was near the laundromat. I checked in with my doctor on plans, and he said they'd reach out if anything changed— otherwise we were still on for the surgery.

Hurricane Katrina made the first landfall in Miami–Dade County of south Florida on the morning of August 25, 2005. The winds screamed and roared as it dropped waves of heavy rainfall onto the city. Palm tree branches and leaves slapped at the windows, creating a constant thrum of noises signaling that the outside chaos was eager to enter our concrete-block home. The power went out early in the storm. Despite the darkness of the storm clouds, we could see the movement of trash and yard decorations tumbling up and down the street through our sturdy hurricane-proof windows.

As Katrina traveled westward toward the gulf, it lost some of its power, but the world would soon learn that it was just getting started on its destructive path. By late afternoon, the winds had died down enough for Marco to walk over to the Bird Bath to check for any damage. The roads were littered with large branches, downed power lines, trash, and even full trees that had been uprooted from the soggy ground with the force of the winds. Walking around after a hurricane can be very dangerous—indeed, this is a period when many injuries occur from un-witting contact with live electric wires and then dehydration and injury during the cleanup process. At the laundromat, he found that the garage-style doors had been damaged by the wind and an inch or two of water had gotten inside. There was no power, so he couldn't use the shop vacuum to clean up, and instead relied on a mop and whatever dry towels he could find.

I was scheduled for the C-section the next day. A phone call from the hospital that evening confirmed my fears that due to the hospital's running on limited generator power, all nonemergency surgeries would need to be rescheduled.

Back at the apartment, Milagros and I worked on preparing a simple dinner of grilled zucchini and canned beans using the propane camp stove.

Without a fan or air-conditioning, the muggy heat was unbearable. I stripped down to a tank top and shorts and impatiently fanned myself while trying to stay distracted with a book. We opened the windows in hopes of a breeze to cool the apartment down. Although we didn't have power for several days, we did have running water. When I couldn't take the heat any longer, I set up candles in the bathroom and stood under a cold shower in the battle with the heat. Never had I craved air-conditioning like this before.

I was cleared for surgery the morning of August 29, four days after the storm had wreaked havoc in our city. I spent the morning in the pre-operative room watching live videos on TV in horror as Katrina made landfall in Louisiana, decimating towns across the gulf, including Daddy's hometown of Biloxi in Mississippi. While the storm raged on along the gulf coast, in Miami, I was wheeled from the pre-op to the operating chamber. Marco walked next to me, but wasn't allowed in until I had the spinal block and urinary catheter inserted. I told the anesthesiologist about my history of back surgeries and the fusion of some of my vertebrae, but I wasn't sure just how far down the fusion went.

I sat on the operating table, and he had me lean forward in an attempt to get a curvature in the spine. The first prick was for a local anesthetic, so I wouldn't feel the pain of the large-bore needle as he tried to push it between my vertebrae into the spinal column. When it was inserted, he would then be able to inject a mix of an opiate and anesthetic (thanks to the opium poppy and coca leaf for these innovations!) into the spinal fluid to bathe the spinal cord in the drugs and achieve a loss of sensation from that point downward in the body.

It soon became apparent that there was a problem. The anesthesiologist continued to push and prod at my spine, attempting to drive the needle in, over and over again. I was scared and in pain. Though I begged for Marco, chanting, "Please," over and over again in supplication, they wouldn't let him into the room. My ob-gyn, who would soon do the surgery, held on to me and gently patted my head as I leaned forward farther as the other doctor tried to find a space in my back to bring the needle through. I'd awoken that morning and done my hair and makeup nicely because I wanted to look pretty for all of the pictures when the baby was born. Lines of mascara were now streaked across my face from my torrent of tears and smeared onto the doctor's surgical scrub shirt from where he held me during the spinal trial.

A decision was made to put me under general anesthesia, and the last thing I remember is panicking as I was told to take a deep breath through the mask that they'd strapped to my face.

Do you have any notions about breastfeeding?" the nurse asked with a thick Cuban accent.

Though it had been hours since I came out of surgery, my head spun, and my vision was still a bit blurry. Her words eluded me amid my morphine fog. I looked up at her, confused.

"Where is he? Can I see him?" I croaked, my throat sore and scratchy from the breathing tube they'd pushed down my throat during the surgery.

"Do you have any notions about breastfeeding?" she repeated. *Notions? What are you talking about?*

"Yes, I want to breastfeed," I groggily answered as my state of anxiety rose to the fore. "Is he okay? I want to see him." Panic set in. *Is he deformed? Why can't I see him? Does he have both legs? Is he missing bones?* Terrifying images raced through my head, the narcotics strengthening the paranoia—the fears of my worst nightmares amplified until they dominated my every thought.

The earliest genetic studies completed on me as a one-year-old predicted that while my parents' risk for recurrence of a disabled child was under 25 percent, my risk for having children with similar birth defects was as high as 50 percent. In college, I had sought out another assessment from the geneticists at Emory, where they had predicted this risk to be closer to 5 percent, based on my constellation of bone defects. We'd tracked the baby's development closely throughout the pregnancy, seeing specialists that could measure his limbs as they developed using advanced ultrasound technologies.

Still, my fears and doubts rose to the surface. Maybe we'd missed something.

I looked to Marco, ready to burst into tears again. I was strapped down with tubes, an IV in my arm, a long catheter snaking up into my bladder. My face was swollen, puffy with the fluids they had pumped into my system. Mascara streaks had puddled under my eyes, making it look like I'd come out of a brawl with black eyes rather than an operating room. My abdomen throbbed where they'd sliced me open like a ripe melon. My back hurt like hell from all of the times they'd tried to push in the needle for the spinal block. Nine angry red holes lined the middle of my back.

"He's fine, everything is fine," Marco said. Mandy and Momma were there, too, nodding their heads in agreement with Marco.

I feared they were trying to protect me from the truth. *They were hiding him from me.*

I looked again at Momma. I knew she would tell me the facts, no matter how hard. I wanted to confirm it with my own eyes. *Why can't I see him?*

DESPITE MY PARANOIA AND WORRIES, he was totally healthy with ten perfectly formed fingers and ten toes, legs of equal length, and no missing bones. Donato Lee was named after his two grandfathers—Donato after Marco's father, and Lee, my father's middle name. In Italian, Donato means "given" or "gift from God." He was a gift indeed.

Once the grogginess of the anesthesia had worn off to the point where I was deemed lucid enough to safely hold him, the nurse brought him in. He had soft brown hair that gathered in a mullet at the base of his head and, at almost nine pounds, was a big boy with a voracious appetite. For some time, we nicknamed him the piccolo vampiro (little vampire) because he nursed so often that it left my nipples raw and bleeding. As a novice to breastfeeding, I hadn't realized that he wasn't latching on correctly, and after finally seeking out help from a breastfeeding consultant, we were able to fix the problem. He was able to get the milk he wanted, and I was saved from the pain.

Although I'd defended my research proposal and passed my PhD qualifying exams just weeks before the birth, I still had some coursework to complete in plant taxonomy. My PhD adviser, Brad—Professor Bennett and I were now on more casual terms—was working with another professor at FIU to co-teach an intensive course on plant identification at Fairchild Garden that fall. Classes were set to begin in early September, the week after I'd returned home from the hospital. We would meet once a week on Fridays at Fairchild Tropical Botanic Garden for a daylong course.

My original intent had been to take the course, but now, dealing with the novelty of being a new mom and recovering from surgery, I was having doubts about my ability to juggle it all. When I called Brad to discuss my reservations, he reassured me, "Cass, parenting is the most important job you'll ever have." He was an incredibly dedicated dad of three girls, and would regale me with stories about their lives. His favorite part of the day was reading the Harry Potter series aloud with his youngest daughter after work. "This class will really help you build those remaining plant ID skills you need for your fieldwork, though. How about if you take a few weeks off, and come when you're ready? Bring the baby with you if you like."

He was offering me the greatest level of flexibility possible in the system we were working in, and I was grateful for that. That flexibility to keep Donato with me to make nursing easy during the full-day class was both liberating and empowering. The course was offered only once every few years, and I needed to hone my skills in taxonomy before I headed into the field alone for my dissertation research in the coming spring. While the prospect of leaping into my research project was daunting, I was also reassured by the tight family support network I had awaiting me in Ginestra. Things would work out.

In those first few months, when Donato wasn't eating, he was often sleeping, and so we were able to do just fine with the class. When he was awake or fussy, Brad and my classmates took turns helping to burp and hold him, and he sopped up all of the extra attention. When I wasn't in class on Fridays, I was able to spend the rest of my time with Donato juggling my computer work from home in preparation for fieldwork.

Little did I know at the time that this experience would portend things to come. Three babies later, all delivered by C-section, and I never had a true maternity leave with any of them—even my last child, who was born during the early phase of my faculty career. I was often back at my computer, or classroom, or in the lab within a few weeks of their births—sometimes still on my pain meds for recovery from the surgery. I built in time at home with them during periods of intensive writing projects—whether grant applications, my dissertation, or research papers—nursing them as I typed, or rocking them in a swing by my desk. This is a loss I will always mourn—that essential time that should be fully dedicated to bonding with your newborn, not juggling work duties at the same time.

This was just one difficult choice among many more to come in my search for balance between my family and my intellectual passions. Becoming a scientist is not a single decision. It is a series of decisions at key turning points in your life. Again and again, you have to ask yourself, *Do I choose science? Do I quit or do I carry on?* The answer for me—so far, at least—has always been yes. Yes to science in spite of the obstacles.

Like so many working parents who play tug-of-war between career and family, progressing in science while also ensuring that my family remains my top priority has become my life's tightrope. I've had to master some acrobatic feats in time management and discipline so as not to fall off the wire, and for the most part, it has worked. Yet, as a field, science must do better. We need to better support parents during this time. Without serious changes in the system, from training to early career development, mothers in STEM in particular will continue to fall behind.

CHAPTER 6

From the Field to the Lab

The scientist does not study nature because it is useful to
do so. He studies it because he takes pleasure in it; and he
takes pleasure in it because it is beautiful. If nature were not
beautiful it would not be worth knowing, and life would
not be worth living.

HENRI POINCARÉ, *SCIENCE AND METHOD*, 1908

In a grassy meadow teeming with wildflowers of brilliant yellow, pale
blue, and pink hues, I brought the wheels of an old Fiat Panda to a
rolling stop. Beyond the pastel meadow, a lush green forest began. It was
the spring of 2006, and we were on family lands at the base of Monte
Vulture, a volcano I knew from previous hikes to mushroom forage with
Marco and his sisters.

I popped the trunk of the old blue car, which I'd borrowed from my
brother-in-law, and, sitting on the bumper, I put on a pair of thick hiking
socks and boots. As with any field expedition, I stuffed some large cloth
laundry bags, gloves, and a set of secateurs into my backpack, along with
my field notebook and GPS. This time, however, I placed some extra
items on top—diapers, wipes, and a baby blanket. I strapped on the baby

carrier, snapping the clips in place over my T-shirt, and turned to retrieve my only company. This wasn't a solo mission.

I unbuckled Donato from the car seat and carefully placed him in the
harness in front of my chest, positioning a floppy hat over his head. I
pulled my pack full of gear onto my back and shut the trunk. After giving
a long yawn, he tried to take in the scene, wide-eyed and curious. At six
months, his depth perception was still developing to a few feet at best, but
he would've certainly been mesmerized by the meadow's sea of colors. I
yawned, too. I was as tired as he was, probably more. While my
C-section incision had healed, I still hadn't recovered my full strength,
and my belly bulged in a muffin-top roll from my favorite pair of field
pants; just below the waistline, the buttons strained against my skin.

I gazed toward the peak. Monte Vulture's days of explosive activity
came more than eight hundred thousand years ago, leaving the surrounding landscape rich in fertile volcanic soil, perfect for the cultivation
of grapevines the region was renowned for—particularly the DOC wine
Aglianico del Vulture. I was interested in a far older crop, though, one
that had been started by the Basilian monks who first settled at the edge
of the twin crater lakes found at the summit. The eighth-century abbey
of St. Michael the Archangel, built into the crater rock, is still there,
managed by Capuchin monks. The ruins of the eleventh-century abbey
of St. Hippolytus are nearby.

Sweet chestnut trees, fruits of the monks' early labors centuries ago,
surrounded us as I hiked up a fern-lined trail. The European or sweet
chestnut (*Castanea sativa*, Fagaceae) is treasured not only as a reliable
source of food—little to no maintenance is required after planting one—
but also as a source of medicine. Each fall, as the air turns crisp, the volcanic forest trees sag with spiky green fruits, which, when opened, reveal
a shiny brown nut. Digging deeper, a creamy light brown cotyledon is
unveiled—packed with vitamins and nutrients, including energy-yielding

starches. Chestnut trees are late bloomers: they can take up to thirty years to start producing fruit. Then again, that's pretty early in their life spans, which can run up to two thousand years.

I sought out a young tree with low-hanging branches, adorned with this spring's pale green catkins that would yellow as the summer progressed. I unpacked my blanket and spread it in a soft clearing. Before getting started on my work, I sat with Donato on the blanket to breastfeed while I reviewed my notes. Local healers had told me they used chestnut leaf teas as rinses to treat skin that was irritated, inflamed, or afflicted with eczema. This was in addition to the historic reports of poultices made of chestnut leaves for skin diseases that I'd read about in old Italian texts. I wanted to include some of these leaves in my dissertation studies on plants used to heal skin diseases.

There was nothing in the existing scientific literature to suggest that chestnut leaves had antimicrobial activities—no extracts had ever been shown to kill or slow the growth of bacteria. Still, local healers swore that it was effective. How could this be? It would have to mean that the chestnut worked against the infection in a way different from what we expected—not killing the bacteria like an antibiotic, but perhaps somehow interrupting its ability to do harm, allowing the immune system to do its own job in conjunction.

In the past few years, there had been a flurry of scientific literature reviews and opinion articles that suggested that an alternative way to battle infection might come from a compound's interference with something known as bacterial quorum sensing. Simply defined, quorum sensing is how bacteria talk to one another and coordinate themselves. They do this by releasing compounds, like little telegraph signals, which gives the bacteria—notice that you rarely hear of bacterium, in the singular, because they always act in a group—their power as a highly invasive army. Some in the field met this idea with excitement, but many were skeptical and continue to be so, arguing that killing the bacteria is the only way to go when battling infection.

The advantages of quorum-sensing inhibitors (or quorum quenchers) are multifold: they can act on their own to reduce the danger of infection by blocking the signals that trigger production of toxic virulence factors that destroy human tissues, or they can be used in combination with existing antibiotics to improve overall outcomes in patients. The problem, though, was that no one had come up with a good drug candidate to act on these pathways. The idea of quorum-sensing interference was purely theoretical. No one had yet found a compound that truly was a quorum quencher for staph. I had a hunch that some of these traditional skin remedies, which didn't kill bacteria but still managed to successfully treat the disease according to local reports, might be the answer we were looking for.

In the most fundamental sense, the question of my dissertation was to test whether or not the ethnobotanical approach to drug discovery actually worked. So, in addition to collecting samples of plants used in traditional medicine for infectious and inflammatory skin diseases (my area of focus), I also collected species used to treat stomachaches, headaches, and diabetes, as well as some plants *without* any locally reported medical applications that I could use as controls. Were medicinal plants really any more pharmacologically active than any other random plant—or was there a placebo effect at play? Upon my return, I would test extracts from all of these plants against staph bacteria, examining their ability to inhibit growth, biofilm, and toxins, and create quorum-sensing interference.

I laid Donato down. While he'd mastered the art of rolling over and even wiggling across the floor like an inchworm, he hadn't yet begun to crawl and wouldn't be able to wander off into the forest while I harvested

chestnut leaves nearby. And with him in the middle of the blanket mur-
muring softly, I got to work.

F IELD RESEARCH IS EXCITING, but it is also physically grueling, mentally
exhausting, and emotionally isolating when you're on your own. Time in
the field is limited, and depending on the scope of the proposed study, it's
often a sprint against the clock from day one. And this hundred-meter
dash is after the marathon of preparation—months of writing research
proposals for approval, submitting grant applications to fund the work,
obtaining the necessary research permits, shipping permits, applying for
a visa for the stay, and setting up on-the-ground and shipping logistics.
Then, despite your best efforts, there are unexpected setbacks. Science
rarely moves forward in a straight line. Whether it's difficulty in finding
your targeted plants in the wild or the right people to interview, or discovering after the fact that your samples didn't dry enough and went
moldy, trouble lurks around every corner. At the same time, it's a total
immersion experience of rough-and-tumble intellectual growth.

I missed working in a team as I had with Andrea and the other visiting scientists in our prior research in the area. I soon realized what a
luxury it was, having him take care of all of the challenging logistics
before I even arrived! Now it was all up to me. Plus, as a new mom, a
baby strapped to my chest (or lying on a blanket, fifteen feet away), the
prospect of fieldwork on my own was daunting. I also missed Marco.
This was our first time apart since we'd married.

But fieldwork was crucial to the successful completion of my dissertation. I'd proposed to document and collect plant specimens that were
used for the treatment of an array of skin and soft-tissue infections under
the paradigm of traditional medicine. After collecting the plants, I

planned to ship them back to the CENaP lab at FIU, where I would extract and test their efficacy against staph. Could these old remedies kill deadly staph bacteria? Could they stop the bacteria from sticking to surfaces in the body, or reduce their power to harm? What secrets would these traditional remedies reveal?

Of course my hunch was that I'd have some luck in answering these questions, but the real challenge was finding the compound. Plants have so many compounds, and the compounds themselves are incredibly structurally complex—with many of them interrelated. It's very challenging to zero in on a single compound, if there is one, that achieves the desired effect. It's similar to the laborious work of a geneticist, trying to locate what one small strip of DNA sequence on which genome might be tied to a certain disease. It involves a lot of methodical searching, trial and error, and many misses.

I planned to interview elders from communities, both Italian and Arbëreshë, spread across the region surrounding Monte Vulture. While the Mediterranean is a biodiversity hotspot with many species having never been investigated for their pharmacological potential, the selection of this field site was strategic. Not only was I familiar with the terrain and communities here, which would ensure I'd be able to complete the interviews and collect the plant specimens I needed, but I also had an incredible support network of family members. Milagros offered to take care of Donato most days while I worked in the field. I could still nurse him on my lunch breaks back in Ginestra.

This choice deviated from what I might have done were I more ablebodied and without an infant. I'd been enthralled by Brad's early ethnobotanical work with the Indigenous Shuar people of eastern Ecuador—their knowledge of the healing and hallucinogenic plants of the Amazonian rain forest was immense. Many students follow in their mentors' footsteps, building on research projects already underway, but after evaluating my options with Marco and my mentors, I concluded that there were too many obstacles for me back in the Amazon. I had legitimate concerns about navigating the rough terrain and thick forest vegetation, across boulder-strewn rivers, and up slick mountain trails—and that was without considering the extra challenges that bringing an infant along would entail.

I quickly learned that this would be the most intense job of my life, requiring extreme discipline and drive if I was going to meet my target of collecting more than a hundred species for my lab studies. My days were filled with plant hunting and sample processing in the mornings, lunch and precious time with Donato at home in the early afternoons, followed by interviews in the late afternoons, and data processing in the evenings. My in-laws gave me space downstairs by the cantina to work, and I set up a small table there. Each day after returning from the forests or meadows with samples, I dumped my haul on the ground by the table and started sorting each bag, one by one. First, I removed any contaminating dirt or other species that had come along for the ride, and then I pressed each sample for herbarium deposit, carefully smashing whole herbs and flowering branches of shrubs between sheets of *Il Fatto Quotidiano*, an Italian daily newspaper, and loading those into the plant press for drying. For bulk specimens, I chopped up the plant tissue of interest—the leaves, fruits, roots—into one-inch pieces. I spent hours clipping away at plants as the radio played Italian pop music in the background. Once the plant bits were chopped, I put them into a heated cabinet Marco and I had constructed. It was basically a Ronco food dehydrator that Marco had built out of plywood and featured wire mesh shelving and a heater with a fan. As the sun set, I'd return to my in-laws' house disheveled, my arms and face covered in dirt smudges, twigs protruding from my short mop of curly blond hair.

It took a day or two for each sample to dry. Once the plant bits were dry enough to snap between my fingers, I packed them in plastic vacuum-sealed bags with sachets of silica gel—just like the ones you find in a pair of new shoes: do not eat!—and labeled each bag with the collection number and date, plant name and part, and weight, and then packed it in a cardboard box with twenty others, and affixed USDA import permits to the side of the box for shipment back to the lab. It was like wrapping a lot of presents for Christmas to send to my faraway family—only I was the sole recipient, and every present was a dead plant!

Andrea's team and I had already found that treatments for dermatological ailments made up roughly one-third of the total plant remedies documented in this region. Other researchers, too, had shown that about one-third of traditional medicines in other areas were applied for the treatment of wound and skin disorders. Yet very few of these had ever been investigated in depth for their pharmacological potential under the lens of Western science. Instead of using a tube of antibiotic cream to treat a scrape or skin infection, village elders turned to rinses, poultices, and ointments made of local plants. This was true even if they had ready access to modern medicines at the local pharmacy or with the village doctor. If someone got cut while working in the fields, they bound the wound with onion skins or stuffed it with the white hemicellulose membranes found at the nodes of freshly cut and split giant reeds. Their whole approach to handling damaged or infected skin relied heavily on natural remedies gathered from their environment.

In total, I ended up conducting in-depth interviews with 112 people, stratified by age and gender, to achieve a balanced view of local knowledge. One by one, I pieced together the details behind the preparation of 38 plant species to yield 116 different topical remedies for skin complaints, ranging from wounds, cuts, burns, rashes, warts, boils, dental abscesses, furuncles, carbuncles, burns, and more. I learned that in addition to using plant materials, animal ingredients were also important.

There were traditions of using products from pigs, slugs, and even humans to create 49 therapeutic formulations. (Garden slugs were plucked from vegetable patches and their yellow mucus rubbed onto warts; the slug would then be hung on a rose thorn—the thought being that as the slug dries out, so would the wart.)

People prepared these plant remedies in many fascinating ways. One preparation method was as an herbal decoction: black horehound (*Ballota nigra*, Lamiaceae) is first boiled in water, then cooled, and used to rinse inflamed skin. For other remedies, the plant parts or resins were also applied directly to the skin, like the milky white latex of fig trees, for treating warts, or mallow leaves, after being heated over a fire, to treat infected boils and furuncles. Some plants were stored in fats to create ointments and liniments, such as the flowering parts of St. John's wort (*Hypericum perforatum*, Hypericaceae), in olive oil for wound healing, or elmleaf blackberry leaves, in aged pig fat for treating carbuncles and abscesses. The leaves were good for treating infection, and the berries were good for eating. The plump fruits were eaten raw or prepared into marmalades—and even the roots of the plant were dug up and boiled for topical applications to treat men with hair loss.

An assortment of minerals also enriched the list of remedies. Overall, out of 165 total remedies documented during these studies, 110 of these had never before been published as part of the traditional pharmacopoeia of southern Italy. I was breaking new ground—and it felt like it. I was ecstatic at what I imagined my contributions might be (once I was able to get back to the CENaP labs and begin to investigate); also my whole body felt like I'd been working twelve-hour days with a jackhammer at a construction site.

Of the plant-based remedies, a few key ingredients stood out. Common mallow is locally known as malva. Zia Elena of Maschito, whom I had interviewed on my first trip to Ginestra, repeated this adage to me:

"La malva, da ogni mal' ti salva," or "the common mallow saves you from every disease." Malva was a panacea—a cure-all.

Decoctions of the aerial parts (flowers, leaves, stems) were taken internally, valued for their restorative properties in the treatment of cold and flu symptoms, and stomachache, and as a postpartum depurative, to help women recover following childbirth. Besides its uses as a tea, it was also applied as a topical rinse for dental decay, tooth and skin abscesses, heat and diaper rashes, bruises, boils, and mastitis.

Like mallow, white horehound (*Marrubium vulgare*, Lamiaceae) was considered a panacea with its own local proverb in southern Italian dialect: "A maruggē, ognē malē struggē." This meant "the white horehound destroys every disease." A stinky member of the mint family locally known as marrubio or maruggē, horehound is boiled and drunk as an expectorant to bring up phlegm and as a liver-protective medicine. Locals also made decoctions of the aerial parts to use as a rinse for athlete's foot, boils, abscesses, cysts, and warts in both humans and their livestock. This would be like a bottle of mouthwash being as effective for menstrual cramps, skin lacerations, and eczema as it is for gingivitis.

While it's amazing how many ailments people treated with either mallow or horehound, the common denominator was that they were mostly used for bacterial infections of the skin and soft tissues. While a dental infection and a boil on your foot seem very different and have their own particular kinds of pain, they're both caused by bacteria. From this fact, I hypothesized that there was some antibacterial activity present because of the plants' complex chemistry.

Of course I'd have to wait until my return to the lab, back in Miami at FIU, to find out.

A month before my return, soon after I'd surpassed the one-hundred-plant mark in my collecting, I got a phone call from Brad. He'd never called me in the field before. His mentorship style was the very hands-off, "push the baby bird out of the nest and let it fly" kind, so to see his number on my phone was very unusual. I wanted to share my good news with him, how well the collection was going, but a shiver of worry shuddered down my spine. I'd just finished feeding Donato and was in the process of walking and burping him, when I picked up the phone.

"Cass—I have some bad news," he began, voice heavy. I held my breath, anxious. He kept me in suspense as he paused. This news was not good, whatever it was going to be. "The training fellowship has been terminated early due to some paperwork mix-ups at the school."

I sat down in the chair with a thud, Donato tight in my arms.

"You'll need to work as a teaching assistant when you get back from the field and we'll try to get you on one of the other school training grants to cover the rest of your stipend," he continued, "but you also need to start thinking about writing some additional grants to support your work."

I felt like the wind had just been knocked out of me—and, along with it, my future. I'd been promised five years of training: here I was in year three and now my meager living stipend of $19,000 a year and my research supply funds had vanished into the hot, dry southern Italian air. I had a month left in the field, and while my in-laws covered my food and lodging, there were other little expenses for Donato's care like diapers and clothes. Plus, I still had to cover the $900 in shipping expenses to send all of the boxes of plants to the USDA quarantine facility in Miami.

Fortunately, Marco's sacrifice in staying behind to run the Bird Bath during our absence had ensured some financial security. By repairing the broken-down washing machines and working long shifts himself, he'd been able to save a few extra thousand dollars to give us a cushion. But it wouldn't be enough to cover my missing income. To continue in the program at FIU, I was going to have to juggle teaching undergraduate microbiology lab classes along with helping out in the Bird Bath and caring for Donato.

I looked at all my plant matter, arrayed messily on the table. Yet I saw the order; I knew which sample needed to go in which bag—mallow there, white horehound here. Donato burped in my lap. It was like the cruel inverse of that moment I'd had in the rain forest, when I stared up at the ojé tree and felt a towering sense of possibility, the branches of knowledge spreading out over my head. Now, everything was within even greater reach, I was further along, I was making progress, but I literally had my hands full. And when I returned to the United States, they would only get fuller—or even tied behind my back. I was exhausted at the mere thought of what was to come.

The pursuit of science is never a straight line—and there's never enough money.

A VISCERAL, OVERWHELMING RELIEF COURSED through my body when I saw Marco's arms, his hair, that smile on the August day Donato and I arrived home to Florida. Marco was just as happy to see me. I ran to him with our little boy strapped to my chest; I forgot all about the dried vomit on my shirt (not Donato's, but rather an unfortunate gift from my seatmate, who spewed on me at the two-hour mark of the seven-hour transatlantic flight). That's how happy I was to see him! And he waited until we were in the car for a while before he ventured, "Honey, I have to tell you. *Tu puzzi!* You don't smell good." That's how happy he was to see me! I smiled to myself. The prospect of a shower, clean clothes, Marco's jokes, and a full night's sleep at home had never been more alluring.

Marco had a way of holding his tongue to protect me. I was grateful for it that day.

The next day, I learned that his tendency to shield me from bad news had been applied in the extreme over the course of the last month. There had been an attempted robbery at the Bird Bath. And this time, it was at gunpoint—and Marco was working the register. He hadn't wanted to worry me while I was away.

We sat together on a hand-me-down floral-printed couch as Donato played with blocks on the fluffy rug near our feet. Marco hit the play button and then wrapped his arm around my shoulders; I leaned in and anxiously watched the video surveillance tape. The timer on the tape showed that it was almost midnight, closing time. There was one customer left folding her laundry, plus Marco and two laundry attendants finishing the last of the wash-and-folds.

Suddenly, a man in a black ski mask strode through the open garage-style doors with a shotgun pointed directly to the service desk, where Marco sat closing out the day's earnings at the cash register. After seeing him, Marco got up from his seat and walked straight up to him, roughly his same height and build, and reached for the shotgun. A struggle ensued, the weapon in both their grasps. With a big backward yank, the assailant wrested the gun back and ran off into the night. The tape showed that it all happened in under a minute. I felt like it unfolded in slow motion, underwater.

"Why did you do that?! Why didn't you just give him the money?" I shouted, leaping up from the couch, angry at him through my fear over what could have happened.

"I thought it was one of my friends playing a prank on me," he explained, "so I just walked up to him and told him to give me the gun before he got hurt."

Marco didn't believe it was a real gun. It wasn't until the entire thing was over, and he returned to his work in closing out the cash register for the night, that the reality of what had happened hit him. His whole body began to shake. He called the police, who came to take Marco's and the other bystanders' statements and made a copy of the surveillance tape. As a concealed-weapons permit holder, Marco had a legal firearm—a pistol—with him at the desk. He never made a reach for it through the whole ordeal.

One of the police officers took him aside to offer some advice as the team wrapped up the police report: "Hey, man. Next time, if this ever happens, you should just shoot him. It's in your rights to shoot him. Trying to stop him without a weapon almost got you killed."

Marco listened in disbelief. He was a long way from Ginestra.

Despite the continued issues with crime in the neighborhood where we lived and operated the laundromat, the option of leaving the business wasn't possible. It was the only stable source of income we had to live on—and, clearly, it wasn't even that stable.

THE DAY AFTER DONATO and I returned, the morning after I watched the video, I got to work on the most urgent of my graduate-study issues: finding funding for my lab supplies and graduate student stipend.

I applied for a Garden Club of America grant to cover the thousands of dollars in chemical and microbiological supplies needed to complete my research project. I began writing my own research training grant application for the National Institutes of Health. I'd tried and failed twice before at securing the F31 predoctoral training award when I'd been looking for funds to cover my schooling in London. I hoped I would be more successful this time. If I could only secure this grant, it would make the last bit of my research training feasible.

As Brad had promised, he found another training grant at FIU to partially support my stipend, and I made up the balance as a graduate teaching assistant, teaching undergraduate microbiology lab classes in the afternoons and evenings. I hadn't fully understood the gift that the fully covered research assistantship had been until then. The fellowship had allowed me the luxury of putting 100 percent into my research and learning. Now, the pace was more intense than ever before, and I struggled to balance the hours I needed in the lab to make progress on my research.

My first box of plant samples arrived at the USDA inspection center by the Miami airport without a problem. After being held in quarantine for a week and inspected to ensure that the samples didn't present a risk of introducing pathogens to US soil, they were released to me to bring back to the lab. It *was* like Christmas!

Even though I'd been the person to collect and painstakingly label, pack, and vacuum seal each sample bag for the shipment, opening up the large cardboard box in the plant-processing lab made me feel giddy. "Hello, elmleaf blackberry!" I'd never been so excited to see dead plants before! "There you are, white horehound, *Marrubium vulgare*, of the Lamiaceae family!" Stack upon stack of bags of chopped-up and dried plant bits awaited their next step in the journey from the wild into my petri dish experiments.

While I had months to wait for responses to my grant inquiries to cover the cost of more supplies—solvents, vials, agar, petri dishes, growth media, reagents, antibiotics, pipette tips, and more—I did have a small stock of supplies I'd bought with my former training fellowship and hoarded before leaving on my fieldwork. It would be enough to get me started on the process for now. First, I needed to transform the bags of plant bits into a powder for extraction.

Next stop: the plant grinder.

The mill was a large, heavy contraption that was capable of taking tough chunks of roots or stems and reducing them into a fine powder. It had been originally designed by the Thomas Scientific company to grind up everything from horse hooves—for glue making—to fertilizer materials in an industrial setting. For scientists interested in plant chemistry, it was like a coffee grinder on steroids: heavy steel blades ripped through the plants fed into the hopper, shooting them out into the collection jar below.

Bedecked in a hazmat suit with goggles and a Darth Vader–esque respirator mask, I spent hours each day in the muggy Miami autumn heat, bathing in my own sweat, as I patiently fed plant bits into the hopper. I felt like a strange mix of Walter White from *Breaking Bad* and the guy feeding Steve Buscemi into the wood chipper at the end of *Fargo*. The mask and body coverings were necessary, as not all of the plants I'd collected were technically safe—neither for skin contact nor inhalation. Every medicine is a poison, after all, and in this case, some from my control group were indeed poisonous. Sample processing was no easy job, and it was the first step of transforming these plant materials into drugs for testing.

The next step required solvents—and the choice of solvent is incredibly important. Plants are transformed into medicines in many different ways. Some are soaked in cold water—like the fragrant herb baths that Don Antonio used in some of his healing ceremonies. Others are decocted, as was popular in Italy, by boiling the plants in water before drinking them as an herbal tea or using the liquid as a body wash. And then in the Peruvian Amazon, I'd learned that steeping the plants in alcohol or spirits was also common, yielding an alcoholic tincture for use. In other cases, as I'd learned from the blackberry leaf and pig fat remedy of southern Italy, a preparation enfleurage, in which the essence of the plant is captured in hot or cold odorless fats, is preferred. All these different methods extract the plant's essence, a beautifully complex array of chemicals that each serve a purpose in enhancing the plant's ability to survive and reproduce.

The choice of solvent matters because it determines which compounds will be drawn from the plant into the solvent, thus influencing both the chemical makeup and the pharmacological activity of the final product. Solvents differ in their ability to diffuse into the plant cells, dissolve the secondary metabolites (defensive plant compounds with potential medicinal activity), and then diffuse back out of the cell. Water is the most widely available solvent—hence its popularity in the preparation of many traditional medicines—but it also is very effective in swelling the plant cells. When combined with an organic solvent, like ethanol or methanol, this allows for greater penetration and increased extraction efficiency of a large diversity of a plant's secondary metabolites.

In my effort to grab the largest diversity of compounds in the plants I studied, I used two methods: the traditional practice preferred by the zie in Italy, boiling the material in water for twenty minutes; and by macerating, or soaking, the materials in aqueous ethanol (water plus alcohol) for three days, much like Don Antonio's rheumatism remedy, a tincture of cat's claw (*Uncaria tomentosa*, Rubiaceae) soaked in rum. I weighed out the dry powder and added the liquid solvent at a ten-to-one ratio.

When the process was complete, I carefully filtered each sample to separate the plant powder from the liquid extract, and subjected all of them to rotary evaporation and freeze-drying, as Tim, the postdoc, had taught me. The result was a single box of small glass vials decorated with labels, starting with "Extract #0001." I felt a sense of great pride looking at that box. I was beginning to build quite a library of extracts for my antimicrobial studies. Each one was filled with crystalline powders in reds, greens, and browns—each representing the unique chemistry of life held within that species. As the chemical library grew, so did my confidence that I just might be able to resolve some of the mysteries of how these medicinal plants worked in treating skin infections.

There was just one problem.

My other shipments, containing most of my fieldwork samples, representing months of grueling work under the hot Italian sun, had not yet arrived and were weeks overdue.

I SAT IN BRAD'S OFFICE, trying my hardest not to cry. I'd never cried in front of a professor, and I *really* didn't want to start now. But I'd reached a breaking point.

Outfitted in his typical Florida garb of shorts and a floral Hawaiian shirt, he nodded his head, intently listening to me explaining my dilemma, pausing every few minutes to spit the juice from the wad of chewing tobacco into a cup, and sipping a soda from another identical cup. Brad reminded me of a Major League Baseball coach—nodding head, darting eyes, mouth full of chew. In the midst of my rant, I wondered, *Does he ever confuse the cups and take a big gulp of his tobacco spit?* Years later, at my graduation, I gifted him with an antique copper spittoon for his office so he wouldn't make that mistake.

"So the plants are lost," he said, part question, part statement.

"Yes," I groaned sorrowfully. I was so full of surging self-pity from the pit of my stomach that I felt I was about to erupt like Monte Vulture into a volcano of tears.

I'd called and corresponded by mail with the Italian post office, the USDA receiving center, and the US postal center—and I'd discovered that the boxes *had* arrived in Miami after all. But, I then found out, they were erroneously sent back to Italy.

"By freight boat across the Atlantic Ocean!" I told him, shaking my head. "Now they can't find any records of them. They don't know where they are!"

I thought about my time away. Had I really lost so many months away from my husband? Had he really lost those early experiences with his son, for nothing? This setback could cost me another year, and I was not interested in lingering in graduate school forever, nor could I bear the thought of another long period of separation. Plus, I didn't have any funding to support another field expedition.

Brad took in my rush of angst-laden words and paused a moment, staring off into space to think. This man, my PhD mentor, had lived among the Shuar, hiked rough mountains, traversed dense rain forest trails crawling with venomous snakes, forded deadly waterways, and survived tropical infectious diseases. If anyone had a solution to an obstacle, he would. I hoped. We sat in silence for a few minutes.

"Well, Cass. This actually reminds me of Rumphius," he began.

"Rumphius?"

"Yes, Rumphius," he affirmed, spitting into his cup.

Brad explained that Georg Eberhard Rumphius was a seventeenth-century German botanist who worked for the Dutch East India Company, best known for writing *Het Amboinsche Kruidboek* (*The Ambonese Herbal*), a collection of six volumes documenting plants from the island of Amboina in present-day Indonesia. It was the first comprehensive work describing the flora of the region, and it's still used as a scholarly reference today. Rumphius spent years cataloging more than twelve hundred species, an incredible task on its own, made even more incredible by the obstacles that he overcame to accomplish it.

Brad leaned forward, took a sip from his other cup, and elaborated.

"So, first Rumphius went blind while collecting and processing the plants," Brad said. "Probably glaucoma. Now, he needed assistants to help him. Next up, his wife and daughter were killed in an earthquake and resulting tsunami on the island. And then, just as he was almost done completing the project, his library was destroyed in a great fire, and all his work went up in smoke!" Brad spat. "Over the next twenty years, he and his assistants remade the entire book! Can you believe it? So, finally, he sends the book back to Europe on a ship, before he's ready to return, and the ship carrying the manuscript was attacked. It sunk, and his book was lost forever once again, this time to Davy Jones's locker. Rumphius and his assistants, working off an older incomplete draft of the book, then took another six years to re-complete it and shipped the copy to the Netherlands successfully." He paused, neither spitting nor drinking. "After all this effort, the Dutch East India Company decided not to publish it initially due to the presence of what they deemed to be sensitive information about the plants of the region. And this great work of his wasn't published for another forty years. Of course, by then, Rumphius was long dead."

I listened in disbelief at the tale. *This* was supposed to uplift and inspire me?

"Listen, Cass," Brad said, spitting into his cup again and then taking a sip of soda. "Rumphius persevered. So will you. Be grateful you don't have it as hard as he did."

This was a textbook definition of tough love. Brad couldn't have picked a worse fate to compare mine to. But again, he reminded me of a baseball coach. He was giving me, his player, a pep talk. *You're in a slump. Get out of it. You can do this.*

I was still near tears as I exited his office, but I realized why Brad's approach still gave me a measure of hope. He was a little like Momma, back when I fell apart in her arms as a little girl, talking about my leg—how I couldn't do things or that other kids made fun of me. Momma never let me believe that I wouldn't succeed. Brad was trying to do the same.

For that, I was grateful.

Although I would've been *more* grateful if he'd simply solved all my problems, and I did, just a little bit, in the heat of the moment, hope that once the door to his office was closed, he'd finally mistake one cup for another and swallow some of that tobacco juice.

Four months later, all my remaining boxes showed up—inexplicably, at the home of my parents-in-law in Ginestra. To safeguard against another shipping disaster, they personally brought them back to Miami during one of their visits to see us. They traveled with the plants as special oversize luggage, tagged with the requisite USDA permits. All of this of course made me think of Rumphius's assistants—they were just as heroic as he was.

While some of my samples had molded over during their gap year of world travel, the majority was thankfully fine, and I went to work with the aid of some undergraduate volunteers to help me make up for this lost time. We made quick progress with the shipment, extracting and moving through the stages of filtration and rotary evaporation of the liquid into a thick tar of plant goo. Next stop: the lyophilizer. In classic Rumphius fashion, it broke. It wouldn't pull a vacuum for the freeze-drying process to work. A new vacuum pump cost thousands of dollars, which was exactly what I didn't have. I asked Marco to come take a look at it to see just how bad the damage was.

Marco came to the lab with Donato. Janna, a new grad student who had joined the lab, was in the student office with her daughter, who was roughly the same age as my boy. Janna needed help with a new experiment, so Brad jumped in to help us both by offering to watch our two toddlers in the office while we worked in the lab. He dragged in a big cardboard box that a piece of lab equipment had been shipped in— instant entertainment for the kids!

Both Janna and I knew just how unusual and special it was to have a kid-friendly adviser. Her former adviser had been the opposite. She'd left her other program after being demoted from the PhD path to a master's when she had informed him of her pregnancy. It was stories like those—all too common in science—that had made me terrified to let Brad know when I was pregnant for the first time. But he'd surprised me with his hearty congratulations and fatherly advice from his experiences raising three girls.

Marco got to work with his tool kit, taking the vacuum pump apart. Inside, he found significant corrosion caused by water and solvents that had been sucked into the lines from improper use in the past. He soaked and scrubbed the pieces, reassembled the pump, and succeeded in restoring the vacuum pressure. It wasn't running at full capacity, but it was good enough to get the job done. That's an accurate assessment of most devices I used in my early lab days.

Working with my undergrad student mentees, we moved from the stages of fully drying down the extracts, recording their dry weight and overall yield from the starting plant material, to dissolving them into dimethyl sulfoxide (DMSO), a colorless liquid that is effective in allowing compounds of different polarities to mix together. This would be essential for dispensing my extract library in tests against the deadly bacteria I was trying to beat: MRSA. It was staph that had nearly killed me as a kid and continued to plague me with skin infections throughout my childhood. This battle was personal.

The next phase of my dissertation was aimed at determining the impact of my extracts—many of which were derived from plants used in the traditional treatment of skin disease—on this superbug's ability to grow, form a biofilm, and release deadly toxins. My training in graduate school had thus far been focused on plant taxonomy, ethnobotanical fieldwork methods, and chemistry. I hadn't had additional coursework in microbiology. To access the additional knowledge I needed, I identified a mentor, Dr. Lisa Plano, an MD-PhD and a specialist in treating new-borns with life-threatening infections, to serve as a key adviser on my dissertation committee. She specialized in MRSA toxins and provided me with valuable training in lab methods and insights on the clinical relevance of the research field. I also had a childhood unusually rich in clinical microbiology experience at hospital laboratories, where I conducted my science fair experiments on multidrug-resistant *E. coli*. I'd learned early on to seek out help where needed and persistently wrote to another expert at the National Institutes of Health who also offered advice on how to troubleshoot my experimental methods to detect how these plant extracts affected the bacteria's behavior and toxicity.

When you're just starting out in science, it's about M&M's: mentors and money. I was well covered with the former, but the lack of the latter continued to dog me. While our lab had the space and infrastructure needed to undertake the research, we didn't have large research grants at the time to support purchasing my much-needed supplies. Losing the training grant had been devastating in terms of losing not only my stipend coverage, but also the few thousand dollars that could have gone toward purchasing materials for my experiments. In stark contrast to the students I train in my lab today, who never have to personally worry about the expense of the supplies that fill the shelves of my labs (I cover those costs with my grant funds), back then, as a student, I was responsible for acquiring every supply, every jar of media, every petri dish I used for my project, with my own research funds. I was thrown into the proving grounds early, forced by necessity to write one grant application after another looking for the much-needed support. I was also responsible for securing my research stipend through either teaching or research assistantships.

Fortunately, I'd been successful in securing fellowships from the Garden Club of America and Botany in Action, but these were quickly consumed by fieldwork expenses I'd incurred and laboratory supplies. No money meant no research progress. I was rejected in my application for the independent training fellowship with the NIH; I took the reviewer critiques to heart and sought out letters from additional external mentors who could help bolster my application and give additional feedback on how to improve it.

After my *fourth* attempt at an NIH individual predoctoral training fellowship—twice when I intended to pursue my degree in London and twice at FIU, since we lost the group training grant—I finally nailed it. This time, the reviewers wrote, "This is an excellent application from an outstanding applicant. Strengths of the proposal include a well-written and well-thought-out research proposal, excellent training program, and institutional environment. A strong point of the application is the mentors and their combined research interests, which perfectly complement those of the applicant. A potential weakness is that the overall project might be perceived as a purely technical exercise with very little mechanistic insight. The applicant has obviously performed a great deal of work in gathering, categorizing and preparing extracts of these plants. However, the final outcome or use of the extracts is not clear."

As I read deeper into their pages of feedback, what did become clear was that they had their doubts about what my research would lead to. These professors didn't share my vision for the pharmacological potential of plants or the value of leveraging traditional medicine to identify specific plants in my research. I shouldn't have been surprised. Drug discovery research on plant natural products had faded out since the 1990s, when attention was shifted to testing huge chemical libraries of man-made compounds. Hundreds of millions of dollars invested into their approach—and they had no new antibiotics to show for it. *Zero*.

Yet, despite these failures, here was this pervasive doubt that anything good would come out of exploring nature's chemistry instead. I felt a mix of frustration and relief. Well, I'd just have to show them where it led. That was the purpose of my dissertation! If I didn't get funding to test my hypothesis, we'd never know whether or not this was a workable model for drug discovery. But the stakes were high, and I believed it was necessary for us to try this path, unorthodox as it may have seemed, to avert looming disaster.

At least the reviewers gave me a score that was enough to put me in the funding pool, and the money came through, giving me the support I needed to complete my dissertation. Now armed with a stipend dedicated to supporting all of my time on research, I was able to phase out of my teaching duties. Plus, it gave me funds needed to buy the $2,000 high-performance liquid chromatography (HPLC) columns needed for the toxin studies and chemical characterization of my most bioactive extracts, as well as some more supplies for my remaining microbiology studies. From that point forward, I realized Brad was right; I had been blessed with much better luck than poor Rumphius.

By the summer of 2007, my lab analyses of the Italian plant extracts were advancing at a rapid pace. I spent long days at work—so long they turned into nights at work—hovering over rows of test tubes and petri dishes. In the microbiology lab, the biosafety cabinet hummed with a constant buzz as the fan ran, creating an air barrier between the workspace where I handled deadly, drug-resistant bacteria and my seat. My pregnant belly pressed uncomfortably against the edge of the biosafety cabinet—affectionately known as the hood—as I stretched to retrieve a test tube hiding in the corner of the work area. Marco and I were expecting our second child, a girl.

Despite the personal costs of carrying a baby, especially for someone with my medical history, Marco and I both yearned for a large family. Maybe it was the time I'd spent with him in Italy, the joy I'd felt amid the boisterous laughter and conversation of the large Sunday family luncheons, tables pushed together and stretched across the living room, full of cousins and aunties and uncles, with children doted on by them all. Perhaps it was also my own roots calling home. With Arcadia being just a three-hour drive from Miami, we had many opportunities to visit. When things weren't too hectic in the lab or at the Bird Bath, Marco and I would load up the car with Donato's playpen and sneak away for the weekend. On chilly winter nights, we warmed ourselves by a campfire and watched the sunset over the small fishpond at Daddy's house, gray tendrils of Spanish moss hanging from thick branches of old live oaks. That time in the countryside with family and friends was like a soothing balm to the hectic pace of city life. It was restorative, and I treasured every visit, every moment.

In front of me, inside the gleaming metal base of the biosafety cabinet, lay a tableau of bacteria, petri dishes, growth media, ninety-six-well microtiter plates, and vials of deep green plant extracts to be delivered to the bacteria. I deftly moved my arm back and forth inside the hood with a pipette in hand, shuttling cloudy liquids teeming with live bacteria into the tiny rows of test tubes that awaited holding staggered doses of plant extracts that I hoped might have some impact on these microscopic nemeses.

I took extra precautions throughout my pregnancy when it came to any of my lab work. I had others handle any potential teratogenic (birth defect–causing) chemicals, and I wore more PPE than usual. I taped my disposable lab coats to my gloves, and remained vigilant in dousing those gloves in ethanol as I worked on experiments with multidrug-resistant staph. Every step I took had to be carefully planned. The experiments involved working with not only antibiotic-resistant isolates of staph, but also strains known to be high biofilm and toxin formers.

Dr. Plano told me about patient cases in the neonatal intensive care unit that she treated or read about in the medical literature, in which infants were affected by hypervirulent strains of staph. Some of those children had scalded skin syndrome, an infectious disease caused by staph that produces exfoliative toxins. They presented with reddened skin at first, followed by painful blistering over large tracts of the body, and then sheets of skin could fall off. The skin is the first barrier the immune system uses to protect the body; without it, these babies were opened up to a horrific suite of other potential infection sources. Plus, skin loss at such a scale could result in depletion of body fluids, causing dehydration and shock similar to that of serious burn patients.

The severity of scalded skin syndrome cases was driven by the capacity of the offending staph strain to produce great quantities of specific toxins. I read and reread those opinion papers on the potential role of quorum-sensing inhibitors. If only we could identify compounds with the ability to block the signaling pathways responsible for the toxin production cascade, maybe—just maybe—the disease could be more effectively treated. I thought about the local uses of white horehound and chestnut leaves as anti-inflammatory rinses and poultices for the skin. If used for skin infections caused by staph in traditional medicine, perhaps they actually did contain the chemistry needed to fight other forms of staph infection, such as scalded skin syndrome. I saw a clear path that my work could lead to in improving the chances of survival and recovery from staph infections like this, and that motivated me to keep pushing forward in my search.

Through all the emotional and physical strain of this period (battling pregnancy fatigue, dealing with the demands of caring for a toddler with a case of the "terrible twos," and continuing my bookkeeping for the laundromat), I learned discipline out of sheer necessity, and because of that, I made progress in my research. I had to finish all my dissertation experiments before I gave birth, and this meant finalizing the tests on hundreds of plant extracts for their activity in blocking the formation of toxins. I'd adapted a method to quantify those toxins to my screening campaign with the extracts—this would be the first time that method was used to hunt for toxin inhibitors. There was no leeway for extra time in the lab after the impending birth. It would be too dangerous to return to a lab full of superbugs with a vulnerable infant at home, especially when my work involved strains that had come from babies who had tragically died from infections caused by those same strains.

Our three-year-old nephew, Trevor, had also come to stay with us for six months, and Marco and I loved and cared for him like we did Donato. Beth was battling dual demons of mental illness and drug addiction and hadn't been able to take care of a child. Trevor had been sickly when he first arrived, underweight, and timid. Fearful of any loud noises, he would cower behind the couch. It broke our hearts. We'd had no idea of the conditions he'd been in, as Beth lived far away in New Hampshire. His overly large clothes reeked of cigarettes and marijuana. We potty-trained him with the motivation of colorful stickers and gently prodded him to eat hearty plates of Italian food each day. We encouraged him to find his confidence in play with other kids in preschool and at the local park we took both boys to in the afternoons. He was smart, already reading, and each evening we read one book after another about Emily Elizabeth and her big red dog, Clifford. With the love and attention he desperately needed, his body filled out and his personality blossomed. He would be reunited with Beth following her treatment in rehab under Momma's gentle care and watchful eyes just before our new baby arrived.

Near the end of my tests, I noted something very interesting—two of the plants that I'd collected significantly reduced the amount of toxin production by staph: chestnut leaves, which I'd so assiduously collected back in Italy when Donato couldn't walk, let alone crawl; and, more to my surprise, the fruits of that pesky Florida weed, the Brazilian pepper-tree. By the time I was set to give birth, I felt hopeful in my work, really hopeful, for the first time in a long while.

Isabella was born on a warm September day in Miami, just two years and two weeks after the birth of her older brother, and this time without the stress and drama of a hurricane. Armed with a stack of X-ray records of my spine, I gave the anesthesiology team the tools needed to ensure that they could find just the spot to push through that spinal block. This time I was awake for the birth, and Marco was there at my side, which meant a lot to me. Marco has been known to faint at the sight of blood, something I of course mercilessly tease him about whenever I can, and there I was on the operating table to be sliced open again. Blood was inevitable. But so was Marco's devotion.

Although the surgical nurses had put up the typical blue cloth tarps at my chest level to prevent us from seeing what was happening down below, my doctor had no idea about Marco's aversion to anything bloody and kept telling Marco to stand up and take photos over the blue tarps with the fancy digital Canon camera I usually reserved for fieldwork. Marco stood up throughout the procedure, snapping high-resolution shots and sitting back down before he passed out.

The end result was a series of moment-by-moment photos of Isabella coming out of my sliced-open belly, covered in gore. He even captured an image of her first cry. A close friend of ours dramatically told us, "I screamed for minutes after looking at those photos!"

My recovery from the surgery was easier this time, and just a couple of weeks after we returned home, it was back to work at the computer for me. I had to analyze the data I'd generated and write papers to complete my dissertation. I also needed to start applying to postdoctoral fellow-ships to continue my training. I was fully supported by my own graduate training grant and was productive working from home, so I didn't have to return to the lab and could focus on my writing.

We established a schedule. Every morning, Marco took Donato to preschool and then went to work at the Bird Bath. I stayed home with Isabella, entertaining her with a bouncy chair and swing; she loved watching me from there. I could type while breastfeeding her at my desk on a U-shaped pillow. Marco and I were a team.

Our system worked, and a year after her birth, by the fall of 2008, I'd successfully published two additional scientific research articles in which I was the lead author. Equally important, I secured a new NIH grant to cover my postdoctoral training expenses with an expert on staphylococ-cal biofilms who was based in Little Rock, Arkansas. By December of that year, I would finally have three long-sought extra letters behind my name: PhD.

CHAPTER 7

Babies and Biofilms

In some Native languages the term for plants
translates to "those who take care of us."

ROBIN WALL KIMMERER, *BRAIDING SWEETGRASS*, 2013

I t's time for the next bag," the nurse said, pulling me out of the World
War II–era love story I'd escaped into and back into the reality of my
gurney. The book had been a welcome gift from my grad school friends
who knew I was miserable in the hospital, although I'm not sure they
fully grasped just how angry I was with my body and its weakness to infection.

It was October of 2008, a little over a year since Isabella had been
born, and I was just a few months shy of graduating with my PhD. Once
again, my stump knee was giving me problems: the skin and tissues
where my calf and lower thigh united had become inflamed and infected.
It was hideous, I knew. Luckily, I couldn't see it. Then again, I didn't
need to. I could feel it, and I knew exactly what was happening down
there, could picture the whole process as if under a microscope, which
was almost as bad. Maybe it was worse.

The nurse swapped out the empty bag of IV fluids that hung next to
my hospital bed for a plump bag with "0.9% Sodium Chloride" stamped

across it. This would deliver the medicine to my body. Then she con-nected the key accessory, a bottle of linezolid, to the network of tubing that ran from the IV pump to my hand, where it was taped in place. We all hoped that linezolid would knock out this infection.

Linezolid is an antibiotic of the oxazolidinone class, effective against MRSA and multidrug-resistant *Streptococcus* infections. It was new to the market and had shown strong activity in soft tissue infections like mine. Following Isabella's birth, firm nodules had formed in the tissue, and they swelled like painful, superstrength balloons; unlike simple boils, they couldn't be lanced. These nodules had truly taken up residence in my body. For the past year, the ebb and flow of inflammation aligned with my menstrual cycle. While I marveled at this synchronicity, it left me stymied by not only monthly abdominal cramps, but also the loss of mobility.

My stump became so swollen every month that I wasn't able to fit it into my prosthetic for at least a week, forcing me to rely first on my crutches and then a wheelchair when I needed to carry anything, whether piles of folded clothes in the laundromat or my growing children. As I was caring for a toddler and an infant at home, I usually relied on the wheelchair: carrying them on one leg with crutches would have taken the sort of acrobatics my body wasn't built for.

AN ACCIDENT IN THE KITCHEN one evening made it clear that something had to be done. Standing in our small galley kitchen on one leg, I propped my hip against the countertop to keep from falling over as I sautéed a tasty mix of onions, zucchini, and tomatoes on the stovetop, stirring them with a broad wooden spoon. My wheelchair was parked just be-yond the baby gate to the kitchen. Bella sat in her bouncy seat, watching me and her older brother, Donato, playing nearby. He'd entered a new phase—one that involved a lot of climbing. I hopped over to the fridge to grab the shredded cheese, distracted for a moment; that's when he managed to scramble up into the seat of my wheelchair.

I dropped the cheese on the countertop and hopped toward the baby gate to put him back to the safety of ground level, but before I could, he lost his balance over the side of the chair and fell to the tile floor. I yanked the baby gate back in my haste to get to him, losing my balance in the process and hitting the floor hard on my left hip. He was screaming and crying in pain, and a big knot had already begun to swell on his forehead. Aching from my own fall, I crawled back into the kitchen to get ice from the freezer to wrap in a kitchen towel. I crawled back on my hands and knees to Donato to ice his forehead; in the meantime, Bella had begun to cry, too. I unhooked her from her bouncy chair, rocking both babies in my arms as I tried to soothe them—and myself—simultaneously.

That's how Marco found us on the kitchen floor when he got home from the Bird Bath, his wife and kids all wailing together. He gently lifted each of us up, and after a good deal of soothing and a walk around the block—Bella in her baby carrier and Donato on his push tricycle— he managed to calm them both down while I iced my hip on the couch. Dinner was ruined, of course. At least I hadn't started a fire. That's what would've happened to Rumphius.

"Don't worry. I'll order some pizza," Marco told me.

This couldn't go on. All that I knew about medicine and infection told me that I needed to take care of this problem. But I'd been in denial. I'd ignored it nine months prior when the swelling first happened, and I wrote it off again six months later when the swelling put me in crutches over the Fourth of July holiday weekend. But the truth was that my stump was becoming infected with greater frequency—and intensity. Still, I'd felt like I was on a roll in my life, personally and professionally, and I so badly hoped that each time was just a fluke. I forced my way through every infection. I'd gone through enough surgeries, too many hospital stays, enough long recoveries while I was a child—wasn't I owed a break now? But I knew things didn't work like that, so I finally resolved to face the problem head-on. Or stump-on.

I tracked down a plastic surgeon skilled in intricate and exploratory surgical procedures that could work in the delicate location to remove the nodules. If something went wrong, if the infection spread, I could lose what little remained of my below-the-knee stump and my knee with it. Below-knee amputees (BKAs) benefit from the mechanical advantage of a natural knee joint—it makes the natural motion of walking much easier than for above-knee amputees (AKAs), who require an artificial mechanical joint adaptor at the knee. While I knew prosthetic technologies had greatly improved since I was a kid, I still deeply feared the potential loss of my knee. Before operating, the surgeon reaffirmed what I already knew: we had to eliminate any trace of the infection first, and that would take some intensive antibiotic therapy to work.

Why do some infections become so persistent? In many cases, biofilms are to blame. Biofilm doesn't refer to a specific type of infection; rather, it is a lifestyle that microbes can assume in various environments, whether on a rock in a stream or a tissue in the body. While most microbes spend time in a state of movement, like bandits roving across the land pillaging, there are environmental conditions that make it much more advantageous for them to hunker down, concentrate, and build a community in one place.

When this happens, they exude sticky substances that allow them to cling to surfaces and one another, forming mushroom-shaped clusters of cells where they can increase their exchange of genes to enhance their survival and slow down their metabolism; this makes them more difficult for antibiotics to target. Indeed, a fully formed biofilm can prove almost entirely impervious to antibiotics and the immune system. An average infection is like a guerrilla attack on the body, or even a large army, where you meet it on a battlefield, outside the walls of your castle: your immune system goes out to fight the marauders. But with a biofilm infection, the invaders not only lay siege to your castle but can even take possession of it—and they start building big mushroom-shaped towers with turrets! With biofilm, they occupy your castle, and your body then needs to try to recapture it—with already depleted resources, and the added challenge of a more formidable and concentrated enemy. As if that weren't enough, biofilms may also involve multiple species of bacteria, and once they're in your castle, they engage in heightened gene exchange when sharing a biofilm architecture, increasing the chances of becoming antibiotic resistant.

I knew this better than most, as I'd been studying this change in bacterial behavior as a part of my doctoral research, investigating how plant compounds change MRSA's capacity to stick to surfaces. The irony wasn't lost on me. While I'd made progress in fighting the bacteria in the lab, it had waged a war on other soil, hitting me at my most vulnerable spot. It likely started as a case of folliculitis (an infected hair follicle) in the tender area where my prosthetic rubbed and irritated the skin in the humid Florida heat. The resulting swelling was a sign of my body's efforts to fight back those impervious cells that clung on to their stronghold in my flesh.

Intensive antibiotic therapy can be effective, but it comes with serious risks. Patients can not only suffer from direct side effects of the drug— ranging from kidney and liver damage to hearing loss—but also experience collateral damage in the gut. Broad-spectrum antibiotics offer the advantage of capturing an array of possible culprits responsible for infection. They round up all the usual suspects. In their sometimes indiscriminate haste, they can also round up and kill the "good guys," or commensals—microbes that coat and fill our bodies and serve essential functions in maintaining our health. Commensals train our immune systems to recognize friend from foe, help break down food, and produce key vitamins for absorption across our gut. They also maintain balance by keeping other, more troublesome microbes in check. Commensals take up some important real estate in the body, and when they are destroyed, other "bad" microbes, like "*C. diff*" (*Clostridium difficile*), can thrive and cause serious disease.

This time I was lucky. I survived this round of antibiotic therapy unscathed, and the stump swelling eventually receded. I had won this battle. But the war wasn't over. I vowed to go back to the lab once I was fully healed to renew my efforts. A few days after my surgery, I was released to go home.

THAT FALL WHILE I recovered from surgery, we prepared for our move to Arkansas, where I had secured a job as a postdoctoral fellow in microbiology to be paid for with my new NIH grant. We sought out a buyer for the Bird Bath. We were on track to pay off the business purchase cost in the monthly installments to my cousin, and with the sale, we'd be able to start the next phase of our life with a nest egg we could use to someday buy a house.

I was on the phone with the new landlord for the building that had housed the Bird Bath—along with a suite of other small businesses: a shipping center, an acupuncture clinic, and a hair salon—for more than thirty years. Her father (the original building owner) had recently passed away and left the real estate to her. And she was not interested in our plans.

"What do you mean, you won't renew the lease?" I asked, exasperated.

"I mean I won't renew it."

"But you know that we have a buyer for the business who has agreed to the terms of your rent increases," I protested.

The accounts were tidy and we had a legitimate buyer ready to move forward. The only problem was that without a multiyear extension of the lease, the business was basically worth nothing. Its value was in the infrastructure (gas and water lines), the equipment, and, most important, the location and customers it served. But those things weren't something that could just be transferred. She knew this and didn't care. She wanted to break up the space into smaller units to make more on rents. She'd also made a number of derogatory comments about our clientele—primarily minorities from the community—making it clear she didn't want "our kind" of customers loitering around the building.

We were pissed off. We were also screwed.

As the end of the year approached, I did what I could to pack up our rental house still wheelchair-bound following my surgery. At one point, our little family was between living arrangements, and all four of us crammed into the back room of the Bird Bath to sleep where we stored the wash-and-fold customers' laundry. During those days, Donato played with toy cars beneath the laundry-folding tables, and I rolled my chair back and forth between machines, swapping out load after load of customer clothes, Bella strapped to me in her infant carrier. At night, Marco and I bathed them in the laundry basin before tucking them into bed on our air mattress on the back room floor.

As the Bird Bath's lease neared its end, Marco worked tirelessly to unbolt and dismantle everything we owned in the laundromat with the help of some friends. We sold what we could to other local laundromats. They were gleeful at the deals they got on the rows of washers and dryers, the change and snack machines, as well as the customers who would soon come their way. Had I been in their shoes, I'd have been gleeful, too.

But *gleeful* wasn't a word in wide circulation in my vocabulary that year. I stared at the balance sheets on the computer screen. We also had to pay off my medical bills from my two hospital stays and the new leg I would need to have built once I healed. Some nights Marco and I couldn't sleep, each of us whispering to the other, "No, it'll be okay, you go to sleep," as we each blinked our itchy eyes, feeling the clutches of debt and uncertainty grasp at our supine bodies like they do to so many Americans well into the night.

This could have ended so differently if the new building owner had only renewed the lease, giving us the security we desperately needed, but I couldn't focus on that. If I did, I'd go mad with rage and desperation. The only thing left to do was to push forward. She never budged, and we lost a lot of money.

Karma played its own role in the end; due to the great recession, the space remained vacant without a tenant to pay the steep rents for more than two years after the laundromat was forced out.

ARKANSAS WINTERS WERE FRIGID. After spending most of my life in the warm bosom of the southeastern United States, I wasn't ready for the snow and ice that greeted us in Little Rock. Marco and I made the drive up from Miami in two days, me driving the kids in our Jeep Compass with just my left leg, and Marco behind the wheel of a twenty-four-foot U-Haul.

I felt guilty about how much he had to do. Still confined to my wheelchair, I was zero help loading and unloading the truck once we arrived. I was only able to keep the kids entertained by racing them around our small apartment on my lap. Conveniently, the apartment was right across the street from the medical campus at the University of Arkansas for

Medical Sciences, or UAMS. We arrived a week before Christmas, and my new job would begin in January 2009.

Dr. Plano, one of my PhD committee advisers, had referred me to Dr. Mark Smeltzer, who would become my new mentor at UAMS. When I'd first reached out to Mark about postdoc job opportunities, he explained that he didn't currently have funding to bring on a new trainee, but if I was successful in finding my own funding, he had room for me. And so I did what I had always done: applied for another grant. This time it was one to focus on studying staphylococcal biofilms. To my mixed delight and surprise, I was successful in getting the grant on the first try.

I'd become more interested in biofilms both from my research and my personal experiences in battling chronic infections. While I'd learned so many new skills in the fields of botany and natural products chemistry in graduate school, my training in microbiology was still limited. Now, as a newly minted PhD, I was ready for the next phase of training. I wanted to take a much deeper dive into the world of microbial pathogenesis— the study of *how* microbes cause disease.

My doctoral studies on plants used in Italian traditional medicine led me to discover anti-biofilm effects in a number of plant extracts. I believed that with more study, refinement, and extraction, they could yield very promising compounds.

Now, it was time for me to begin microbiology boot camp.

Mark's labs were divided into three rooms with one long research bench in the center of each. Glass flasks used in the shaking incubators lined his many shelves. In these flasks, his team grew vats of staph bacteria in broth. They deployed these cloudy white bacterial cultures in experiments to investigate the genes responsible for the sticky behaviors of staph in plastic micro–test tube plates and even in segments of IV catheters like those that had fed my body the antibiotics in my recent hospital stay. They developed tools using the microbes, removing certain genes to examine how that affected their behavior in different environments. They also made some of the staph strains glow in the dark, releasing luminescent signals when certain genes were activated.

Over the next two years, I threw myself into the work, leveraging the new tools I was learning about to examine the biofilm-blocking possibilities of the plants I'd collected—especially any infections linked to the skin and soft tissues. Using the luminescent and fluorescent reporter strains of staph, I explored the various genes that the extracts might be targeting. I learned how to use human blood plasma to form a more robust biofilm for testing my extracts, and while my preliminary studies had led me to develop my research proposal on the evaluation of white horehound extracts, the new work I was doing in Mark's lab led me to another, more potent extract—that of the elmleaf blackberry.

My fieldwork in Italy had revealed that locals used skin rinses of horehound decoctions for various inflammatory conditions, and they'd also noted the specific use of blackberry leaves formulated with animal fat in treating skin infections—abscesses, furuncles, and carbuncles. All these conditions involved biofilm. Abscesses are surprisingly common; in 2005, more than 3.2 million people were treated for abscess infections in US emergency rooms. Local infections trigger skin abscesses and typically involve the accumulation of pus and liquid within the dermis (the layer of tissue below the epidermis, which contains nerve endings, blood capillaries, hair follicles, and sweat glands). Severe abscesses have to be surgically drained and treated with antibiotics. Back when I was a high schooler in the OR, I'd observed doctors drain a massive abscess in the thigh of a morbidly obese patient. When they'd sliced into the abscess pocket, more than a gallon's worth of fluids came out. I struggled to hold back from gagging behind my mask: the smell was a grotesque blend of decay and rot.

In addition to working with Mark, I met Dr. Cesar Compadre, another faculty member at UAMS, who enriched my training in natural products chemistry. This included setting up many experiments involving open column chromatography, in which large glass tubes, three inches in diameter and two feet in length, were loaded with a silica slurry. The active blackberry extract, 220D, was loaded on top, and over a couple of days, series of different solvents selected for differences in their polarities were poured through the column to isolate different fractions for further bioactivity testing against staph biofilm models. This process eventually led me to home in on fraction two, named 220D-F2, as the most potent inhibitor identified in the mix.

This fraction was exceptionally good at preventing the bacteria from sticking to the human-plasma-coated surfaces. While this showed excellent potential as a prophylactic agent, I didn't know yet if it could it also remove an already formed biofilm. In other words, did it have therapeutic potential?

Besides their role in abscesses, biofilms are also the major culprits in medical device infections. More than a million hospital-associated infections are due to bacteria growing on medical devices implanted in the body. This was yet another kind of infection I knew all too well. At the age of twelve, I developed scoliosis. The doctors put me in a nighttime back brace, fearing that a full-time one would be too psychologically damaging for me. Every night, I strapped on the brace that aimed to twist my spine in the opposite direction of the S-shaped curve. Due to my orthodontic issues of a recessed upper jaw, I also wore a football helmet that served as the outer anchor to the colorful rubber bands that latched onto my braces. Despite the months of hot, uncomfortable sleep in the contraption, my spine continued to curve until it reached a debilitating curvature that had begun to crush my internal organs. The option of last resort was spinal surgery, and we pursued it. Dr. Price sliced my back open from my neck to my bottom and screwed in metal rods to straighten the spine as much as possible, reaching a thirty-two-degree upper/twenty-three-degree lower curve—a great improvement over the more than fifty-degree curve I'd reached.

The sharp pains began almost two years after the rods were first implanted. My body was rejecting them—something that happens to only a tiny percent of patients. Rejection of medical devices due to inflammation and infection at the site is sometimes caused by bacterial biofilms that have latched onto the foreign material—reaching the site via circulation in the bloodstream following something as simple as a minor scratch. There was no other recourse than to remove them through surgery and dose me up with antibiotics. As my spine had successfully fused, even with the rods removed, I would be safe from development of a more severe curve.

My hope was that 220D-F2 would be helpful in treating cases like mine. Unfortunately, after many rounds of tests, I proved that 220D-F2 showed no benefit over the growth control in removing the biofilm once it was already established. This was very disappointing. But I also noted that antibiotics weren't very effective either. It took ten times the normal dose of antibiotics to even make a small dent in the number of biofilm-associated cells. So neither one was very effective in a therapeutic setting. Then I had an idea. I combined them. In test after test, when I employed 220D-F2 with traditional antibiotics, they were effective in fighting biofilm, yielding drastic drops in the number of live cells sticking to the IV tubing.

This was, really, my first major scientific breakthrough. In this case, neither the magic bullet antibiotic of Western medicine nor the less powerful approach of traditional medicine worked on its own, but they made one hell of a team! In fact, the findings were so promising that UAMS filed a patent to protect the innovation in the hope of this leading to development of a drug product to help patients battling staph biofilm infections.

I was committed to moving this work forward, hopefully someday to the clinic. I took advantage of some local courses in entrepreneurship on how to write small-business grant proposals, known as Small Business Innovation Research grants, to the NIH. I reached out to an old friend from college, Sahil Patel, who'd gone on to business school. We'd remained tight friends over the years, our respective families included in a group of six couples that gathered for summer vacations. An avid guitarist, he provided the tunes for our vacation sing-alongs. We formed a small business, which would license my university patent for work toward commercialization. We called the company PhytoTEK LLC—*phyto* for plants and TEK for technology. But TEK also stood for something else: *traditional ecological knowledge.*

I'd finally found a successful way to marry what was best about both worlds of medicine, my dream ever since staring up at the ojé tree in the Amazon. But this wasn't an ending. It was only the beginning.

WHILE I WAS RELISHING the progress I was making at the lab bench, I was struggling to fit in with my peers. Working in a basic sciences department was a new experience for me, and I found it difficult to connect with my lab mates. I was an outsider from day one. These other scientists didn't feel like *my* people.

My desire to take a multidisciplinary approach to nearly every question I investigated made me a very odd duck in a traditional scientific environment. I wanted to appreciate the magic behind ritual healing while also investigating the chemical mechanisms by which the plant ingredients in such processes worked. My lab mates wanted to understand the purpose of what each gene in the microbe served. To be clear, translational science is built on the foundations of basic science; my work wouldn't be possible without those studies and mechanistic insights they illuminate. It's just that my mind didn't work that way.

But I also quickly discovered how much my approach differed from that of antibiotic drug–discovery scientists. Other scientists developed specialized cellular assays to screen libraries of lab-made compounds in search of drug leads. To me, though, those chemical libraries were irritating and boring. I wasn't interested in compounds with little structural diversity made out of the minds of humans—I wanted to explore the compounds made by nature, a domain still largely untouched by even the most intrepid of scientific explorers.

There was really only one group of people who understood me. Ever since that first ethnobiology conference in Athens, Georgia, where I'd met Andrea, I knew I'd *found my people*. Throughout graduate school, Brad had encouraged my attendance at the annual meetings of the Society for Economic Botany (SEB); there, I began to form lifelong connections with friends and colleagues, ethnobotanists and ethnobiologists who dedicated their careers to investigating the myriad and often mysterious ways that humankind connected to nature for survival and for the enrichment of life in art, music, medicine, and more. It was through those early experiences that I learned about the Open Science Network (OSN) in Ethnobiology, a National Science Foundation–funded initiative to train upcoming science educators and develop curricula to be shared on an open and accessible platform on the web, which was a novel concept at the time.

The OSN held educator-training workshops at the SEB meetings, and I learned from incredibly experienced educators as well as the young cohort members who participated in the network. I owe much of my current teaching skills to this training period. It certainly hadn't come from my brief stint as a middle-school teacher or even from graduate school, where I'd been taught only the most basic pedagogical techniques. Back then I was more worried about a student accidentally blowing up the teaching lab than whether or not I had provided a detailed and useful rubric. To be fair, my worries about lab explosions weren't unfounded. One student did in fact catch on fire by spilling ethanol on their glove and getting too close to the Bunsen burner, while another had attempted to light the lab gas directly from the gas outlet, risking some major fireworks in the building had I not lunged to turn the switch off in time.

Whenever I reunited with fellow ethnobotanists scattered across the world at annual meetings, they gave me a renewed sense of purpose and the drive to push on. Plus, there was PhytoTEK, which represented the ways that I wanted to do science—drawing upon traditional wisdom of nature's therapeutic potential and uniting that with cutting-edge modern scientific techniques in microbiology and chemistry. Sahil and I worked toward building the company through developing our business plan and even submitted applications to start-up competitions. Because we were based in different states (me in Arkansas, Sahil in Maryland), we were emailing and on the phone all the time, especially in the run-up to our first start-up competition. I was convinced we could take matters into our own hands and push new therapies for battling staph, like 220D-F2 from blackberry roots, closer to pharmacy shelves. And, who knows, maybe we'd be able to make a living out of doing it.

WHAT YEAR DID you graduate?"

And by that he meant *from Harvard*. Sitting in a classroom with stadium-style seating in Washington, DC, I waited for the judges to find their seats. Together with Sahil, who was a Harvard Business School alum, I had entered the school's alumni business plan competition hoping to gain some support for PhytoTEK LLC. And now, a large man in a suit with a crew cut was sizing up the competition by cutting me down.

"I didn't," I replied. "My business partner did."

Nodding as if he'd known this was the case all along, the man launched into a rousing encomium of himself and his company—it was going to be a game changer, a major disruptor, a driver of innovation. Of course he didn't want to brag or anything, but he'd already "taken meetings" with some "very high-level" people who really "knew the space." The message he thought he was sending me was clear: out of the six teams competing, his was the best. I was no threat.

When he walked away to sing his own praises to someone else, I wrung my hands nervously. Even nincompoops can faze you. I kept going over the pitch in my head. Sahil and I had practiced the seven-minute presentation endlessly, recording ourselves on our cell phones, checking the times for the transition from one presentation slide to the next. It had been a long and painful exercise, and if we hadn't been friends since our freshman year of college, I might have hated him for it. In this moment, though, I was grateful for his insistence on the intense practice sessions. At least I felt more prepared. This wasn't something we could just wing.

I looked around the room. Rows and rows of white men in suits. That's when I realized it: I was the only woman in the competition—and the only person without a business degree—and Sahil was the only person of color. I was there to explain the science, while Sahil would handle the business end of things.

I'd been in this situation a handful of times—feeling totally inadequate and insecure before a presentation, only to be further intimidated by a competitor—and it never got any easier. The first time was when I was in sixth grade, and I competed in my first State Science and Engineering Fair (SSEF) of Florida. I felt a sense of wonder and amazement unlike anything I'd ever experienced before as I walked down the rows upon rows of display boards in the convention hall. Some of the biggest competitors at the Florida SSEF each year came from Brevard County: they were the children of NASA scientists and engineers, or students who benefited from the advanced programs at those schools. Their displays included TV video monitors and robotics demos.

My display was pretty simple. Personal computing was still new back then, and while I typed and printed my project display elements (Purpose, Hypothesis, Results, Conclusion), I didn't have access to graphing software. I hand colored my title banner and drew my bar graphs using colored pencils. I wore my best Sunday church dress for the big day, my bangs blown out into a curled pouf that sat on my forehead, as was the style of the 1980s.

"Your board looks awesome," I earnestly told the boy in the booth next to mine as he stood there, arms crossed, in khaki pants, a button-down shirt, and tie. He nodded and looked over at my display.

"That's so lame! Why didn't you use a computer to make your graphs?"

I didn't know how to respond and turned back to my board, trying to hold back the tears. I was embarrassed and hurt, and later that day as I walked away from that competition with no prize in hand, a part of me agreed that he was right. My display board was lame.

Now looking around the stale DC room as an adult, I had a sinking feeling that I was surrounded once again by khaki-pants boys, all grown up.

Finally, it was our turn. Sahil and I walked to the stage.

The front screen in the room gleamed with our green company logo, and our tagline, "Eliminating Infection. Saving Lives," was stamped across the bottom. Sahil launched into the introduction.

"PhytoTEK supplies a unique biologic solution for the infection-control market, which is a large and rapidly growing $3 billion global market that cuts across industries with many applications," Sahil said smoothly. "PhytoTEK focuses on staph-related infection, and the key markets that the management team is addressing include human healthcare, food processing, dental care, and veterinary care. Each of these currently have significant pain points caused by staph infection."

I continued, "For example, human healthcare-implanted medical devices, such as a knee or hip replacement, can become infected by staph not only during surgery, but even years following surgery. When this happens, often the only recourse is a period of intensive antibiotic therapy followed by surgical removal of the device—which can mean an invasive orthopedic surgery—followed by debridement of the infected tissue, more antibiotics delivered through a temporary device, and then another surgery later on to replace it with a long-term device."

I paused to make eye contact with each judge and then explained, "Once you've experienced one of these infections, the likelihood of the replacement becoming infected again is as high as 40 percent for some medical devices. Imagine for a moment the pain and suffering these patients go through. It's more than just a statistic, and it's no wonder that some with chronic infections of the knees prefer to just have the leg amputated than deal with the treatment plan again and again. We need a better solution. I think that combination therapies and medical-device coatings using both antibiotics and biofilm inhibitors like ours could be it."

All our practice and training kicked in—the flow of our presentation was polished and professional. We went on to explain that we formed PhytoTEK to hold the license to my first patent filed at UAMS during my postdoctoral training period. The technology was a standardized botanical extract derived from blackberry roots that was very effective at blocking how staph bacteria stick to surfaces, including medical devices.

I felt good after our pitch, but my confidence was still shaken by the sheer quality of the other presenters. Besides other medical technology start-ups, there were also several information tech platforms, and promising business concepts focused on software and internet apps that could be rapidly launched. Our model, on the other hand, required years of R & D and substantial investment to gather the necessary data for regulatory approvals through the botanical drug pathway.

Antimicrobial research and development is not a speedy process. It can take ten to twenty years for a truly novel class of antibiotic to go from molecule to approved product, and the average cost to first approval is around $1.3 billion. Then there is an estimated $350 million in additional costs linked to the supply chain and pharmacovigilance (tracking any adverse events in reaction to the drug) that are incurred for the first ten years after it is approved and brought to market. The poor economic market for antimicrobial agents is at odds with their sheer necessity to combat disease. These are *not* small hurdles, and while all can agree that new drugs to combat resistant infections are needed, these issues have led to big pharma largely exiting the antibiotics research and development space.

After the last team wrapped up their pitch, the judges excused themselves to another room to deliberate. Time crawled by, and we joined the other teams in selecting some snacks from the buffet arrangement as we waited. Some of the other presenters crowded around the crew-cut man—my khaki boy all grown-up—and it was clear everyone thought his team was the frontrunner. I was too nervous to make much chitchat and busied myself by aimlessly scrolling through my phone.

At last, the judges returned to announce the results. I put my phone away out of respect. I guess I was finally going to learn the name of Mr. Crew Cut Khaki. After announcing third place and then second (which Mr. Crew Cut Khaki received), I was slack-jawed when the judges announced Sahil Patel and Cassandra Quave as the first-prize winners! I was in utter shock. This was huge. In addition to the bragging rights and a huge affirmation of what we were trying to do, we got very helpful feedback on our written business plan and the pitch, and we advanced to compete at the Harvard Business School Club of New York, where thirty teams applied with their plans, and six had the chance to deliver their pitches in the semifinals.

Having altered our business plan and fleshed out our presentation to better address some of the market hurdles, we sat in a large room on the twentieth floor of a Manhattan high-rise building a month later. This time, there was no crew-cut khaki kid there to throw me off my game, but it was still all incredibly intimidating. This was only my second trip to New York City, and the energy of everything from the chatter in the high-rise conference room to the city lights and noises—taxi horns and ambulance sirens—left me in a daze. Our repeated practice sessions and prior success in DC gave me the confidence to join Sahil at the front of the room to deliver my part of the pitch.

To our delight (and my surprise), we won that competition, too, and advanced to the global finals in Boston, where we competed against teams from all over the world. I was battling some serious impostor syndrome in that room of champions, baffled at our success, but exceedingly grateful nonetheless. Ultimately, our passion had shone through, as had our careful planning. A perk of winning the New York City competition included coaching from a business leader in a major sector of pharma in preparation for the next challenge to come on the Harvard campus.

In Boston, competitors flew in from South Africa, Singapore, Europe, and all across the United States. We continued our practice sessions over the phone and in person once we reached Boston. This would be the big prize—the winner would take home a package that included high-quality legal and accounting services from a top global intellectual property firm plus $30,000 to use for the business activities.

The day of the pitch, I paced in my hotel room, running the presentation over aloud, gesticulating with my hands for emphasis as I do only when speaking Italian. We'd refined the written plan with our advisers, polished the pitch, and knew our roles.

After we presented, the judges posed questions about how we planned to fundraise to support the next series of animal tests needed to validate the technology. UAMS had a biomedical ventures incubator just across the street from the main medical campus, including state-of-the-art laboratories, a conference room, office space, and access to the core research facilities for chemistry, microscopy, and animal studies. Our plan was for me to transition from my job as a postdoc to go full-time in running the science end of the company in the venture park, establishing us as a brick-and-mortar instead of virtual company. To avoid losing equity early on to investors, we planned to bootstrap our product R & D and seek out nondilutive funding through the NIH small-business research grant mechanisms.

Ultimately, our pitch wasn't enough to carry the day, and we came in second, as the runner-up. It was a huge honor, really. But at the same time, honor wasn't money—and money was the *real* thing we needed. Science or business, business or science—money was what we needed to push either forward.

Straight nines? Are you kidding me?" I asked aloud in dismay as I looked over the results at home over breakfast. I leaned over our kitchen table to wipe up a puddle of milk that Donato had spilled while eating his cereal, and then scooped some more chopped-up fruit onto Bella's high chair tray.

Marco refilled my cup of coffee and sat down next to me to look at my laptop screen. "What does that even mean?" he asked. We were looking at the grant review results from one of several applications I had submitted to the NIH on behalf of PhytoTEK.

"Well, a score of one is perfection . . . and a score of nine in any category is basically the worst possible score you can get. I didn't even know reviewers gave scores that bad," I huffed as I scrolled the report down the screen.

Pointing at the computer, I continued, "Look! This reviewer even wrote their opinion in all capital letters, like they're yelling at me! What a jerk!"

While not all of the reviewers gave such harsh scores (or ALL CAPS notes), the review panel reached a consensus: I was too junior to take on this initiative, and they *really* didn't like the use of a botanical extract composed of different chemical entities. In their opinion, I needed to focus on single compounds, preferably those that killed the bacteria like a normal antibiotic. They didn't see the utility of targeting biofilms.

Marco rubbed my shoulders and kissed me on top of my head.

"It'll get better," he said.

But would it? I doubted that it would.

Despite early momentum with the competitions and my publication on the biofilm-busting technology, I was making no progress in my pursuit of grant funding for the company. The economy had been in a bad spot ever since the market crash of 2009, which coincided directly with our move to Arkansas. Marco, who initially had a hard time finding work, eventually got a minimum-wage job for a tree-trimming service picking up branches, and later found somewhat better work as a maintenance repairman at an apartment complex. It was a big change and was not as personally fulfilling to him as running the Bird Bath, where he had been his own boss (as well as the boss of others). But still, we shared the perspective of doing whatever it took to stay afloat and were grateful for the work.

After we lost the Bird Bath, the combination of his work and my limited postdoc income wasn't enough to cover my medical expenses for the new prosthetic leg and day care for the kids. I took on a second job teaching microbiology lectures and labs to undergraduates at night at the University of Arkansas at Little Rock (UALR) as an adjunct professor. The position didn't pay much, but it kept us out of a financial hole. These grants and the company were supposed to be our way out of this struggle to make ends meet. I'd been so confident that this would be my launch moment. I thought everyone would see the obvious value of what I'd discovered. But it wasn't that easy or simple. I should've known that this, too, would take as much work as everything else. Still, it stung. I continued to fail with my grant applications, and the time on my NIH postdoctoral fellowship was running out. Soon, I'd be out of a job.

CHAPTER 8

A Lab of My Own

The pathway to discovery originates right here, right now, in
this very time and place, in the fields, forests, and outdoor
habitats that have sustained our species since the beginning.

JUSTIN M. NOLAN, *WILD HARVEST IN THE HEARTLAND*, 2007

With my NIH fellowship ending soon, and things with PhytoTEK
failing to develop as Sahil and I had hoped, I was losing some of
my faith. I continued to hit dead ends on grant applications. The reviewer comments were nothing if not consistent—I needed to focus on
single compounds; they didn't like my anti-biofilm or antivirulence strategy; they didn't feel our company possessed the right infrastructure to do
what we proposed. If only I could just get situated in a stronger research
environment than my start-up, I felt, then I might have a chance at gaining some much-needed momentum. I still believed in my vision for the
ethnobotanical approach to drug discovery, but it seemed that no one
else shared my belief.

Owing to Momma's old advice that I should always keep in touch
with people who had helped me, telling them of my progress (or what
I felt to be my lack of it), I sent an email in May of 2011 to one of my

mentors, Dr. Lampl, at Emory. She wrote back quickly and asked to schedule a time to talk.

Dr. Lampl told me about a new program she was establishing at Emory, the Center for the Study of Human Health (CSHH). It would focus on the science of health, covering the topic in the most holistic way possible. The goal was to unite experts in the health sciences from across campus—linking public health, the humanities, neuroscience, nutrition, and more, all with the goal of designing coursework that explores human health from many angles.

"Are you interested in coming on board as a teaching postdoc to design and teach courses for the center?" she asked.

This was exactly what I needed! There were no job offerings for me at UAMS once my NIH funds were gone. My night job of teaching microbiology to undergrads at UALR as an adjunct wouldn't come close to paying my share of our household bills. My many attempts at grants to get PhytoTEK off the ground had not been successful. Atlanta shone out to me like a beacon of light and hope, where opportunities abounded, along with a paycheck.

There were so many other intellectual perks as well. A new center with a multidisciplinary approach: here, I could pursue my work, surrounded by other people asking big questions, including experts questioning the way their fields were organized. Plus, I'd missed the rich environment that a liberal arts university offers. As an undergraduate, while very focused on studies in the sciences, I'd loved the opportunities to engage in the arts by attending poetry readings, museum expositions, concerts, and theater performances on campus. Graduate school and my postdoctoral training had been largely devoid of this enrichment, and the opportunity to get back into a high-tier research and arts community was too good to pass up. So I decided to head back to Emory.

Finances, again, stood in the way of an easy move. In addition to the kids, Granny (in her late eighties at the time) had also come to live with us in Arkansas. We'd bought a small home in an estate sale and couldn't afford both the rent in Atlanta and the mortgage payments in Little Rock if Marco left his job. So Marco stayed to keep working, while we packed up a U-Haul and the kids and I made the drive to Atlanta alone. We spent the rest of the money we had for Granny to fly the following week because the drive would be too hard on her. We were blessed with the help of friends and connections from back home who'd moved to Georgia, and they came to help me carry our furniture, clothes, toys, and kitchen goods upstairs to our second-floor two-bedroom apartment.

I enrolled Donato in kindergarten and Bella in day care, and launched myself into the new work with fervor. That first fall semester, I designed three new courses, learned how to teach the existing ones already on the books, and juggled single parenting and caregiving for my two little ones and Granny. At night after dinner, dishes, laundry, and bath time, I tucked the kids into bed and continued to teach microbiology as an adjunct at UALR online through a virtual platform. I missed being in the lab, but teaching was rewarding, and I loved the energy of the center where we were marshaling all of our creative ideas toward building an enticing educational program. It was so hard to be away from Marco—and hard for him to be away from his children and me—but thankfully by Christmas he was finally able to join us, having secured a job with an apartment complex in Atlanta.

In 2010, THE NATIONAL CENTER for Complementary and Integrative Health (NCCIH) at the US NIH issued a special call for mechanism of action studies on botanicals (basically, studies that uncover how compounds from plants work on the body or against a disease); the R01 research grants would be worth nearly $2 million for each recipient.

Working with Dr. Alex Horswill, a microbiologist and expert in staph communication pathways I'd met at the gram-positive pathogens conference in Nebraska the year before, he and I submitted a collaborative application under the auspices of PhytoTEK as the host organization. We proposed to further investigate work I'd begun as a graduate student—not on the blackberry, but rather on the chestnut. It was chestnut leaves I'd collected in the mountains of Ginestra when Donato was just an infant, still unable to crawl. In the lab, chestnut leaves had proved to be potent quorum-sensing inhibitors for staph infections. By focusing on the chestnut, we would expand the scope of PhytoTEK's technologies to encompass botanicals that targeted bacterial biofilms (blackberry) and toxins (chestnut).

As I explained earlier, *quorum sensing* describes how bacteria communicate with one another and modulate their behavior based on those communication signals. This process is critical to the infection cycle. On its own, a bacterial cell (a single bacterium) can't do much damage; it's when many cells get together as a group and coordinate behavior like a well-trained army that they can cause serious problems when they attack your body. *Quorum quenching*, on the other hand, describes the blockade of that signaling system.

Staph infections are wily. They release toxins that pop red blood cells, like a sharp dart tossed at a water-filled balloon. These toxins ravage cells and tissues, leaving painful necrotic lesions—zombie skin—in their wake. Even worse for us, staph have the unique ability to persist inside *and* outside our cells. When immune-cell responders capture staph, like when Pac-Man gobbles up dots, those bacterial cells that should be killed in the process fight back by releasing a toxin bomb *in* the immune cell—forcefully blowing it up! Indeed, staph bacteria have even been found to release signals into the environment of the body to lure immune responders to them and, like a Trojan horse, attack when they've made their way inside.

Through their unique weaponry, staph have become incredibly effective at destroying human tissue to generate new food for their growing population, but they've also devised ways to fight back the immune system response by directly attacking those cells as well. The suite of toxins produced by these pathogens are responsible for a vast array of infections in the body, all determined by body site. The toxins ravage bone tissues in osteomyelitis, spread through the body in bloodstream infections (bacteremia), and lead to one of the greatest killers in the world today—sepsis.

My hope was that by studying chestnut we might find an effective quorum quencher, giving us a whole new weapon in the fight against staph. There are no drugs on the pharmaceutical market that specifically target quorum sensing or this toxin-production pathway in bacteria. And so there was no established path to drug approval to follow for a therapeutic approach like this. Trying to use compounds in chestnut to fight staph would certainly be considered an "out-of-the-box" approach to treating infection.

But I'm a firm believer that we can't solve big problems by thinking in terms of "the box." Whether you're inside or outside, the box is a construction that limits our imagination. The box is a barrier to creative thought and innovation. Right now, when drug developers think about how to discover or design new antibiotics, they always do so within the box of classic antibiotics (e.g., penicillin) as the central framework of their design. I don't think that's sufficient anymore. But even an outside-the-box approach is flawed, too, like searching for new versions of antibiotics or remodeling chemical scaffolds of existing antibiotics, because the *sole* goal is to kill the microbes. This paradigm blinds us from seeing or discovering other potential pathways that have not yet even crossed the chasm of human imagination.

Throughout the course of history, major paradigm shifts have been required to advance the ways that humans not only understand the scientific workings of the natural world, but also make major advancements in the field of medicine. There are plenty of examples of this in the field of microbiology. I once had an opportunity to visit the Museum Boerhaave, a museum of the history of science and medicine that holds a collection of Antonie van Leeuwenhoek's original microscopes. A Dutch lens crafter, Van Leeuwenhoek was the first human to ever see and document living microorganisms—which he referred to as "animalcules" as a way of describing the tiny animal-like creatures he found in drops of water and other sources. I noted how very different these early microscopes, organized on display behind protective layers of glass, were from today's tools, in both shape and size. I wondered at their simplicity and elegance, composed of a small thin metal plate roughly the width of my thumb, with screws and brackets holding bits in place. The specimen was mounted on a needle-point tip and could be viewed through a tiny, expertly ground lens. This innovation allowed Van Leeuwenhoek to change the way that humans viewed the world! He was the first person to ever see red blood cells, bacteria, muscle fibers, and spermatozoa.

I imagined what it would've been like to be a scientist in the seventeenth and eighteenth centuries—the thrill of discovering new forms of life and expanding knowledge for all of humanity tempered only by the loneliness of drawing and speaking of things that most could hardly fathom, and still others would refuse to believe. Van Leeuwenhoek's famous 1676 paper known as "Letter on the Protozoa" gave the first detailed account of single-celled organisms living in different environments, with astonishing resolutions at sizes as small as one micron—half the thickness of a red blood cell! And yet other scientists in the centuries after him were unable to replicate his designs with the same levels of accuracy, thus doubting and even dismissing his discoveries. Their lack of ability and understanding held back advances in the fields of biology, especially microbiology and our understanding of infectious diseases. It wasn't until the nineteenth century when scientists like Louis Pasteur and Robert Koch began to form an understanding of just how important microbes were to medicine and that there was a causal link between certain microbes and specific diseases.

I felt a kindred connection with this figure and others who approached science from a different angle. Yet their tales were cautionary ones: they broke away from the status quo with such ideas, much to the consternation of their colleagues and peers. How many scientists had made breakthroughs in the past, only to be overlooked, mocked, or even ostracized in their lifetimes? Only the most fortunate were lauded centuries later, like the Hungarian physician-scientist Ignaz Semmelweis, who in the nineteenth century was ridiculed by his peers due to his urgings to wash their hands between performing autopsies of women who'd died from childbed fever (a bacterial infection) and delivering babies in healthy women. Had they followed his advice, the lives of countless mothers would have been spared. Today, his tale, and that of Joseph Lister, the British surgeon and "father of modern surgery" who advocated for the use of sterile surgical techniques, reverberate not only in the classrooms of modern schools of medicine, but also with the broader public.

Advances in analytical chemistry techniques have revealed new ways to interpret the chemical signals of plants. *This*, the twenty-first century, is the century that will be noted in history books as the point at which scientists learned to read the language of nature. And what an opportunity it is! However, will humans take advantage of these new tools?

What if we're facing a microscopic enemy that we can't simply kill with the nuclear bomb of antibiotics? What if what we need to look for is not a classic antibiotic but a compound or constellation of compounds that target host immunity instead? What if a way to fight resistant infections is something we haven't been focusing on, like shutting down the microbe's ability to do harm to the body in the first place? Or what if a chemically complex natural extract, in conjunction with a classic antibiotic, might be the real key to ensuring or prolonging our survival? After all, is the goal of medicine to kill the microbe or heal the patient?

Our biggest limitations are in the variety of biological assays and the types of questions these assays enable us to ask in the lab. The study of natural products has the potential to open up radically new areas of scientific inquiry, leading us to understand health and infection from novel perspectives. To find the big answers, though, we have to ask the right questions. To do that, we must create space for the mind to work where the box does not exist. *There is no box.*

W<small>HERE</small> I <small>WORKED</small> <small>BEFORE</small>, someone did the dishes for us," the young man said.

I forced a neutral expression as I nodded my head, struggling to contain my disbelief—this was the fifth research technician candidate I'd interviewed to work in my newly formed laboratory. I'd made it clear that it would be just me and the person I was hiring who would be working in the lab for now. Did he think I'd be endlessly cleaning his Erlenmeyer flasks for him? I wrapped the interview and waited for the next applicant.

I knew how important this first hire was to the successful launch of the lab. The previous candidates included a mix of people with limited lab-science experience and even one audacious chem-lab supply salesman who thought it would be a good chance to pitch new equipment to me.

It was July 2012, and a typically hot summer for Atlanta, humid and in the nineties. I sat on one of the wobbly chairs in the air-conditioned lab, the best chairs I could find that weren't completely broken. The lab was a good size, just under one thousand square feet, featuring two long black-topped benches down the middle, and two chemical fume hoods at one end, where we could work with noxious chemicals. Those were its strong points. Electronic junk lined the countertops, old keyboards and a mix of Apple computers with slots for floppy disks that were cutting edge—forty years ago. The room was dusty, with water stains on the ceiling tiles from roof leaks, and dingy gray walls that hadn't seen a fresh coat of paint since the Carter administration. The sinks stank with something that must have fermented and festered in the U-traps over the years, and the wall cabinets were filled to the brim with old samples of monkey shit. No joke. The lab was literally filled with samples of dried feces collected from primates by the professor who used this space before me. No one had bothered to clean things out before I took over the space.

Although Alex and I hadn't been successful in our first attempt at landing the big grant, initially earning an impact score of 52, when I submitted the revised application under the auspices of Emory instead of PhytoTEK, we nailed it with a score of 30, the magic number that put us in the running for funds!

Dr. Lampl had been a huge help in this process. She introduced me to Dr. Dennis Liotta, a chemistry professor at Emory who, along with pediatrics professor Dr. Raymond Schinazi and researcher Woo-Baeg Choi, co-developed the molecules 3TC and FTC, which revolutionized AIDS treatment through their activity in preventing or delaying HIV from replicating and infecting other cells. Dr. Liotta graciously offered to share his expertise and my application as a co-investigator, and along with the promise of the lab space from Emory College (should I be successful in getting the grant) and the exceptional research infrastructure that Emory had to offer, I passed one of the toughest NIH reviewer gauntlets!

With research funds in hand, I was officially a principal investigator (PI) and was able to vault out of my role as a postdoc and into a new job title of visiting assistant professor at Emory while I searched for a tenure-track option at Emory and elsewhere. I had the same heavy teaching load, but it came with a small salary bump and my much-needed lab space to undertake the research that I'd proposed in the grant: chestnuts. All I needed was a hardworking and intelligent research assistant. That's when Kate walked through my door.

Of course! One of fourteen brave students who'd shown up twice a week at 8:00 a.m. for my Botanical Medicine and Health class, Kate was a high performer. The class was the first that I'd designed and taught at Emory, a challenging senior-level course that included presentations and a plant monograph paper, peer-reviewed by other students in class. Kate earned good marks in that class and in others that I taught. She'd just graduated with her bachelor's degree and was looking for a job. Kate was a perfect fit.

Tall with straight mahogany hair and a laugh that lit up the room (or the monkey-shit-infested lab), Kate was a hard worker, having waitressed through college. I could teach her what she needed to learn in lab methods, and she wouldn't mind helping me clean the place up.

We found all sorts of weird things during The Purge—a tiny mouse-size guillotine, reams of heat-sensing paper from old-school printers, and boxes upon boxes of old electrical wiring and plugs. I sorted through anything valuable that remained—beakers, flasks, spare equipment parts we might be able to repurpose, and specimen bottles and other bits of glassware that we were able to clean up. Although I was bringing in substantial overhead funds to the university with my grant, my temporary position as visiting assistant professor didn't come with any lab start-up dollars.

While $2 million sounds like a lot of money, roughly $650,000 of that went to university overhead costs, and of what remained, nearly half went to Alex—my collaborator in Iowa who was to perform the mechanistic and animal studies. My role was to lead the project as well as handle the chemistry and pharmacology aims. The remainder would be divided up into chunks over five years to pay research staff salaries and purchase consumable lab supplies. That didn't leave much for me in direct funds to set up my lab, and there were a lot of chemistry tools I needed: the most expensive was a high-performance liquid chromatography system. An HPLC would allow me to analyze and separate compounds found in my plant extracts. But I also needed a rotary evaporator, ultracold freezers, and a lyophilizer (that freeze-drying machine I used at FIU). Altogether, those items totaled more than $100,000, leaving me with very little to cover all of the other basic things needed to explore plant chemistry.

So Kate and I got creative.

We scavenged the rest from other labs that were closing down. Whenever a lab was set to be decommissioned, word came out on university emails. We monitored those notices like hawks. Although most of the good stuff—the pipettes, vortex genies, and other functioning lab equipment—was often already cleaned out by others from that lab's academic department, we were able to liberate some useful things. The key was to get there fast, and there were many days when we'd drop everything mid-experiment to run across campus with the rolling lab cart in tow. It was like *Supermarket Sweep*! Upon arriving, we'd split up and rush to the shelves and cupboards for glassware, old lab coats, boxes of gloves, pipette tips, petri dishes, test tube racks, and more, loading up the cart as quickly as we could with anything that could be cleaned or repaired. I took the lab coats and washed them in bleach a couple of times, then hand sewed on any missing buttons. Using my personal funds, I allowed myself the small pleasure of taking them to the embroidery shop in a strip mall to get "Quave Lab" added in cursive script stitched with emerald-green thread above the pockets.

In the chemistry building, there was a regular junk pile in the hallway where other labs tossed out broken equipment. Anytime Kate and I went on a supply run to purchase more chemicals there, we grabbed anything that could prove useful—dumpster diving to nab broken centrifuges, vortex genies, sonicators, heating water baths—and brought them back to the lab. Of course neither of us had a clue how to repair these things. However, I knew someone who did.

After spending a full day working maintenance at the apartment complex, Marco came to the lab at night to tinker with our finds. Sometimes it was as simple as replacing an electrical cord or a blown fuse to get a piece of equipment back in action; other times Marco took apart large machines screw by screw, cleaning and reassembling the small motors. All in all, we saved tens of thousands of dollars by dumpster diving and fixing things up. Those early lessons I'd had as girl watching Daddy strip down and rebuild heavy equipment engines using spare parts found at the junkyard had come in handy. It's like Bocephus (aka Hank Williams Jr.) sang: "A country boy can survive" (and so can a country girl!).

WHILE THINGS HAD BEGUN to progress well in the lab for me and Kate, there was one major core service that the university was lacking to support my research efforts: a herbarium. When I asked around, I was told there was a herbarium, although no one knew where it was or who was in charge of it. I reached out to the head of the biology department at Emory to find out how to resurrect the collection.

The chair of the biology department, a tall man with an impressively large gray mustache and a gruff demeanor, towered over me as he fumbled for the keys to the herbarium workroom. There was no label or signage on the door to indicate that there was anything particularly interesting about the room. It was just one of many anonymous doors on the long hallway where undergraduate biology labs were taught. It could easily have been the entry to a janitorial closet.

When we finally walked inside, a musty odor of dust and preserved plants hit me like the unique perfume of an old museum. The room was a cluttered mess, with stacks of broken chairs from the teaching laboratories and a random hodgepodge of empty bookcases and desks. Twenty years of persistent neglect had rendered it on par with Pompeii, frozen in time and buried in dust. A decrepit freezer chugged with effort in the corner; I worried it was about to explode.

"This is the workroom," he announced. Then he walked back out into the hallway and unlocked the door to the collection room, where the plant specimen sheets were stored in large metal cabinets. "The rest of the plants are in here," he said. I grabbed the handle to one of the cabinets to take a peek inside. It fell off.

I tried the next one, which opened with a bit of force, revealing strips of fabric lining the door. "I wonder if those are mercury lined," I said aloud, just stopping myself before my hand touched it in my exploration of the inner contents of the cabinet. The shelves were bursting with specimen folders, packed in tightly—too tightly—and that made me worry about whether they were damaged.

"Yeah, it's probably full of something poisonous to keep the bugs out," he responded.

"Do you have any plans for the collection?" I asked, and then explained that I needed a functioning collection for my research.

"I almost gave it away last year," he responded. "The University of Georgia was willing to take it off our hands. There aren't any botanists here anymore. Nobody uses it."

"I could take it on as curator," I offered, fully expecting him to say no.

"Yeah, sure, why not. There's no money for it, though, so you'll have to sort that out on your own."

And that is how I became the director and curator of the Emory Herbarium—a collection no one wanted or used and that had no funds to support its operations. Fabulous.

When people hear the term *herbarium*, visions of lush tropical greenhouses often come to mind, but an herbarium is more like a library—instead of shelves of books, there are stacks of flattened plant specimens that have been dried and glued to sheets of acid-free paper. In other words, herbaria are a type of natural history collection, and each herbarium sheet represents the state at which a plant was found in a certain point in time in a certain geographic space. Herbarium specimens serve as a record of life and are essential to research across the fields of botany, ecology, climate change research, ethnobotany, and, in my case, medicine.

As with the lab, I cleaned out the junk and scavenged for chairs and furniture that weren't falling apart or could at the very least be repaired. Records were sparse, and the cluttered filing cabinets offered little information. Without going through each cabinet of specimens and sorting through thousands of fragile specimen sheets one by one, I couldn't even be sure of how many specimens were in the collection, how many species were represented, or if we had any explicit examples of species upon which the description and name were based. An adjunct professor in environmental sciences who had been around for years mentioned that there could be some important historic collections there from the southern part of the state, but that was all of the information I could find. Digging around the university archives in the library, I learned that the herbarium had been founded by the head of biology in 1949, with his wife serving as the first collections manager. I had a feeling that this collection had a rich history and perhaps even some secret treasures to be found. To learn more, I needed to find a PhD botanist with a strong background in plant systematics and a love of collections to help me.

But first I needed to raise some money to pay for this hire.

After spending days digging around in herbarium cabinets, I found a special collection of plant specimens that held deep ties to Georgia's history. Working with the philanthropy fundraising team at Emory, we had a breakthrough with a philanthropic donor who was willing to support the work with a gift of $250,000, but who wished to remain anonymous. My pitch was to undertake a total revitalization of the collection, with an emphasis on its historical relevance. The plan was ambitious—and would take years—but it was the only path I could envision in which these and the more than twenty thousand specimens could be preserved for the long haul. This would involve updating the specimen management plan with new environmental controls—installing a room dehumidifier and dropping the temperature to keep the specimens cool and dry, repairing the metal cabinets and removing the poisonous mercury lining, and shifting to a safer pest-control product. Errant insects could ravage a collection like this, and we were lucky that it hadn't suffered water and pest damage over the twenty years it had been abandoned.

The National Science Foundation supported a training program, iDigBio, which was dedicated to the digitization (photography and data entry) of natural history collections, including herbaria. I took advantage of a number of training sessions offered through iDigBio at the University of Georgia and Valdosta State University. It was so incredibly helpful to connect with other curators and learn all of the best practices for care of the collection. I didn't need to reinvent the wheel—I just needed to learn how to do things right.

I proposed to hand repair, annotate, and digitize every single specimen in the collection, one by one—a seemingly insurmountable task at the time. Using the funds from the philanthropic donation, I hired PhD botanist Dr. Tharanga Samarakoon to oversee the day-to-day work. She was the perfect person for the job, skilled in taxonomic methods, with an eye for detail and a great rapport with students and trainees.

Then I recruited as many volunteers as I could find from my classes—and even the larger community—to dedicate ten to fifteen hours per week in the collection. They were trained in how to carefully affix specimens to the acid-free paper, enter the location, year, plant name, and collector name into an online database, and then take high-resolution digital images of each herbarium sheet, all of which was loaded to the Southeast Regional Network of Expertise and Collections (SERNEC) website. Once this process was complete, anyone from anywhere in the world would be able to use the collection in teaching through the web portal. If someone was looking for a particular plant species that we had in our collection—maybe a rare endemic one from Italy, or even one now endangered in the United States—they could log on and find it with the click of a few buttons! Our high-resolution images, coupled with the annotated data entered on its date and location of collection, made it a high-value resource for both research and education. It was like our very own digitized Library of Alexandria for plants!

BACK IN THE LAB in the fall of 2013, I sat at the high-top lab bench, busily crunching numbers on my laptop in a massive Excel sheet that contained the molecular masses of hundreds of compounds found in extract 224C, the ethyl acetate partition of sweet chestnut leaves that we'd been examining. Over the past few years, my collaborator Alex and I had made progress in examining the ability of chestnut leaf extracts to interfere with quorum sensing in staph cells, effectively blocking their ability to release tissue-destroying toxins.

All plants are incredibly complex, and chestnut is no different. My task was to find what compound or compounds were responsible for the quorum-quenching properties I'd already proved that chestnuts exhibited, like looking for a needle in a haystack. I was sorting through data on the chemical masses of each molecule in the mixture, representing their precise size; this was based on the number of carbon, nitrogen, oxygen, hydrogen, and other atoms that each unique molecule was composed of. The tools of modern mass spectrometry enabled us to get the precise masses of exceptionally tiny molecules, and even from the state of an extremely complex chemical mixture. I could use these measurements to find putative matches against known compounds in chemical databases. This was all part of the process of uncovering the chemical makeup of this medicinal plant.

In addition to my big NIH grant on chestnut, I'd been successful in landing a couple of smaller grants that enabled the postdoc hire and the purchase of more lab supplies. And so Dr. James Lyles, an avid fisherman with a love of the outdoors and plant natural products, became the first postdoc to join my small research group. While I was on the computer, James taught Kate how to set up a two-foot-tall silica (silicon dioxide) gel column in one of the chemical fume hoods. Besides James, Kate, and me, the research group had grown over the past year to include even a few undergraduate students who'd taken my courses and volunteered their time as interns in the lab to gain more experience in the field. James fitted the long glass column in place with clamps and ran a slurry of silica in liquid until it formed an evenly distributed packing of wet, solid white powder along the length of the column. Next, he added a dark green powder, composed of the chestnut extract that had been bound to the silica, on top. As different solvents—selected for their varying polarities— were poured on top of the green powder, colored bands migrated to the bottom, each band representing different groups of compounds that we collected as the liquid dripped out in a steady, but slow, stream.

I'd just come back from a recent collection trip in Italy, where I'd shipped box loads of dried leaves packed into vacuum-sealed bags to the lab (this time they didn't take an around-the-world gap year). We'd ground the leaves into a fine powder, steeped them in alcohol, filtered them, and dried down the dark green liquid to make extract 224, the 224th extract created from plants collected in my growing chemical library that we stored in the freezer. At each level of cleanup and testing, we moved closer to identifying the compounds responsible for the effects observed in our cellular models for MRSA quorum-sensing behavior.

Work wasn't going as quickly as I'd hoped, though. And I had a lot of ideas that I wanted to explore further with the other plants in my collection beyond the chestnut. What I really needed was to establish a space to do our own microbiological studies, instead of being wholly reliant on collaborators. That work required very different equipment and environmental controls than the phytochemistry, or plant chemistry, lab had.

I'd been persistent in applying for jobs, mainly targeting ads focused on antimicrobial resistance—tenure-track assistant professorships in microbiology across the United States. To my great dismay, I didn't get any callbacks—no stream of eager invitations for interviews by phone or in person. The Center for the Study of Human Health, where I was currently employed, wasn't a department and thus couldn't offer any tenure-track positions. To stay, I'd need to find a department tenure home on campus, and maintain a split appointment with them.

What had I done wrong? Surely there must have been some massive error made when I'd applied. After all, I had the unique prestige of a top research award that most people earned only after years of being established in a professorship with all of the support of a tenure-track position, such as hundreds of thousands of dollars of start-up funds to buy necessary equipment, graduate students in the lab, reduced teaching loads, and peer mentors. The majority of other NIH R01 recipients had all of the advantages in getting their awards, and I had none. My success was the result of doggedly scrapping my way forward. That NIH grant should have been my golden ticket, but I was *different* and my area of research was *unusual*—I didn't fit the checkboxes for the expertise that search committees were, well, searching for. I wasn't a classic choice, and there was nothing that I could do to fix that. So maybe I hadn't done anything wrong. I just wasn't right for what people were looking for.

And yet, as had happened so many times before, a path forward appeared from an unexpected place. In addition to teaching and working in the lab, I'd taken advantage of a business and technology boot camp offered at Emory aimed at supporting faculty and researchers that shared an interest in entrepreneurial steps for their science. My hope was that the course would help bolster my progress with getting PhytoTEK off the ground. One of the other members of the boot camp was Dr. Jack Arbiser, an MD-PhD dermatologist with an entrepreneurial spirit and a love of natural products. We were kindred spirits. After I ran into him on campus one day, he told me, "I have someone I want you to meet."

A few weeks later, I walked from the lab across campus to a large medical building on Clifton Road, just down the street from the main hospital complex. To my great surprise—though I tried to hide it well— that meeting was no simple meet and greet. Rather, I'd walked into a full-on recruitment pre-interview with Drs. Robert Swerlick and Suephy Chen, the chair and vice-chair of dermatology at Emory, respectively, with no prior preparation. I was dressed casually in jeans and T-shirt with sneakers; they wore suits.

Once I got over my nervous reaction to the situation, we eased into a great conversation. They were curious about my research; I launched into an enthusiastic description of our hunt for quorum-sensing inhibitors hidden in chestnut leaves. I felt myself settle back into the grooves of pitching that Sahil had helped drill into me.

"Our studies have already shown that the refined chestnut leaf extract, 224C-F2, shows a dose-dependent response in lowering signaling behavior in all four known accessory gene regulator types—and it does this in the absence of growth inhibition. This means it can shut down virulence factor production across all staph strains, regardless of their antibiotic-resistance profiles," I explained to them.

I sketched out a model of how the signaling system worked on paper, explaining that through blocking that process, I could effectively trick the cells into "thinking" they were alone, instead of surrounded by their brethren. "If they can't sense one another, they behave differently. And that is to our advantage. Think about it like this: if the staph cells were a dog, 224C-F2 takes the teeth out of the dog's bite. No teeth, no problem. The cells may be there in the body, but they can't attack or defend themselves. This could make them much more vulnerable to the immune system or treatment with antibiotics."

"Have you given much thought to the types of diseases this innovation could someday treat?" Dr. Chen asked.

"Yes, staph is involved in so many different types of toxin-driven infections. Just thinking about the skin and soft tissues, there are many options: abscesses, scalded skin syndrome, infected wounds, necrotizing fasciitis, and more."

"Have you thought about atopic dermatitis?" Dr. Swerlick asked. "Staph is known to heavily colonize atopic lesions in patients."

I knew the very basics about atopic dermatitis, or eczema, but I wasn't fully acquainted with the level of staph involvement in the disease. I was intrigued. "I haven't thought much about that possibility as a disease to target," I responded, knowing how to pivot. "But I'd love to learn more about it."

At one point in the meeting, they asked, "What's your number?" I froze. I realized in that panicked moment that I had no idea what I should even be asking for. Did they mean salary? Or laboratory start-up package, or both? I did know from my bit of business training to never be the first one to say a number, and so I didn't speak right away, eventually mumbling that I'd get back to them.

They were looking to expand the laboratory research arm of the department, which was primarily clinical focused, and liked the translational potential of my work for research on the skin microbiome and associated diseases like eczema and acne. I'd long been interested in the topical use of herbal medicines for skin and soft tissue infections—that's what most of the remedies the zie in Italy taught me about. All this time, I'd been so focused on the microbiology end of things I'd never even thought to explore job possibilities in the realm of dermatology. This would allow me to build collaborations with clinical faculty and take a closer look at the host—or skin—side of things when assessing my plant extracts for their healing properties.

"Would you be interested in coming back for a formal interview? This would involve giving a department seminar and interviews with the department faculty," Dr. Chen inquired.

The more I thought about it, the more I liked the idea and the translational potential this path just might hold.

NINE MONTHS LATER, in November 2013, I tossed my car keys on the countertop and kicked off my black wedge dress shoes as I came through the door. I'd exchanged my typical lab attire of jeans and T-shirt for a black pencil skirt and a button-down business shirt for work. Marco was there in the kitchen, layering thinly sliced zucchini, shredded carrots, mozzarella, and tomato sauce for his delicious vegetarian lasagna. Lucio Battisti crooned sweet 1970s Italian tunes from the portable speaker.

"How was your first day at work?" he asked.

"Well, there were *a lot* of assholes," I said. He poured me a glass of red Sangiovese wine and passed it to me after I plopped onto the barstool in the corner, exhausted.

"But I thought you really liked the department faculty," he responded.

"No, I do love the faculty. They're awesome and welcoming! Plus, Kate and I are making great progress in setting up the microbiology lab. I meant it literally. There was presentation slide after slide of images of skin diseases of the anal region at grand rounds this morning," I explained.

All the dermatology faculty and medical residents were expected to attend this meeting several times per month, when medical experts presented case studies and data on different topics in the field of dermatology. Though I'd been with the college as a visiting assistant professor and attended faculty meetings and research seminars for two years now, this had been my first seminar as a member of the School of Medicine faculty. Those were just a lot of graphic medical images of diseases that occur in skin folds of the nether regions to take in that early in the morning while munching on the complimentary bagel and fruit salad.

After battling some serious impostor syndrome wondering how I'd wound up in this seminar room with the new job title of assistant professor of dermatology and human health, I began to appreciate the ways that the medical residents described the various colors, forms, shapes, and textures of the skin lesions. They used words that reminded me of the afternoons of botany lessons where I'd examined the shape, texture, and color of various leaves and flowers in the muggy Miami heat. Same principle and terminology, just different surface to explore—skin, not leaves. I was giddy with the possibilities that could come of new research collaborations in the department.

The deal I'd brokered over months of negotiations between the college and the School of Medicine was a split appointment. I'd teach two undergraduate courses in the college with the Center for the Study of Human Health, and raise research grant money to cover three-quarters of the medical school portion of my salary. The position came with start-up funds to buy equipment and supplies and a bump in salary. Most important, it came with the additional lab space I desperately needed to launch the microbiology and human cell culture experiments to fully examine the pharmacological potential of my plant extracts. Now I had a lab on each side of campus, one we nicknamed Phytochem (short for the phytochemistry space where we studied the chemical makeup of bio-active plant extracts and were certified as a USDA-approved quarantine facility, allowing the import of plant specimens from around the world), and the other nicknamed Micro (short for the microbiology space where we tested those compounds both against multidrug-resistant infectious agents and for safety against human skin cells). The challenge of the deal—which I didn't fully appreciate at the time—was that I was getting the same teaching load as other college science professors with the added pressure of being required to cover a big chunk of my salary with grant funding, like med-school science professors. In other words, I was getting the best and the worst of both worlds. Most important to me, though, I finally had the three core pieces of my research program in place—a phytochemistry lab, a microbiology lab, and the herbarium—all of which were key to my mission and vision to build a leading research group to investigate the pharmacological potential of plants used in traditional medicine for infectious and inflammatory diseases.

It was an unusual path, and to survive, I knew I'd have to swim fast or I'd sink like that ship carrying Rumphius's manuscript.

PART III

Medicine

Castanea sativa

CHAPTER 9

The Sea Cabbage

I have frequently been questioned, especially by women,
of how I could reconcile family life with a scientific career.
Well, it has not been easy.

MARIE CURIE, *AUTOBIOGRAPHICAL NOTES*, 1923

In July 2017, I stood by the small port on a square concrete dock just
large enough for the bulky car-laden ferryboat to slip in and out in the
morning and afternoon. The crystalline blue water sparkled under the
rays of the bright morning sun. White-and-turquoise fishing boats
bobbed up and down with the gentle roll of small waves in the Mediterranean Sea. Rows of white houses decorated with brilliant blue trim
much like the boats joined as a united wall along the small stone-paved
village streets. I slipped on my sunglasses and tilted my head up to admire the mountain that dominated the island, the base of which the village abutted.

I was on the Italian island of Marettimo, situated west of the city of
Trapani in Sicily. Joining me was Marco and our three kids; Akram, one
of my pharmacology graduate students from Emory; and two local collaborators from Palermo, Dr. Alessandro Saitta, a mycologist and an expert of the fungi of Europe, and Dr. Alfonso La Rosa, a naturalist and an

expert in the flora of Sicily. This island jewel of the Mediterranean is a quiet place for Italians to vacation for the summer while enjoying nature hikes and the sea. For others, it's where they grew up, raised their families, and continue to live. For me, it was the third and last stop on an international team expedition I was leading to discover and collect wild plants to fuel my lab's antibiotic-discovery pipeline. There was something on top of the island's mountain that I needed, and I was determined to get it.

The Mediterranean basin is just one of thirty-six global hot spots of biodiversity; other examples include Madagascar and the Indian Ocean islands, Polynesia, the east Melanesian islands, Caucasus, the tropical Andes, the Himalayas, New Zealand, and the horn of South Africa. A biodiversity hot spot is defined as the earth's most biologically rich, and yet threatened, region of terrestrial life. To be a hot spot, the region must have at least 1,500 endemic species of plants found nowhere else on earth and to have lost at least 70 percent of its native vegetation (often due to human development). An incredible 44 percent of all plants on earth are confined to these hot spots, comprising just 2.3 percent of earth's land surface. Though my colleagues delight in teasing me about my choices of field sites—*who wouldn't want to spend their summer on magical islands in the Mediterranean?*—in truth, the rationale for these choices was based on where I could find the greatest variety of useful plants and, in turn, the greatest number of potentially novel and useful structural chemical scaffolds for our drug-discovery pipeline. Little by little, by either leading expeditions or developing international collaborations, my team has added samples from eight of the thirty-six hot spots to my growing collection of plant extracts.

We're in a race to collect and study these potentially priceless resources before they are lost forever due to the rising pressures of climate change, habitat loss, and overharvesting for herb trade networks. As of January 2021, the global population is 7.8 billion people, and roughly 80 percent of them, or 6.2 billion, live in economically underdeveloped countries. Medicinal plants constitute the primary pharmacopoeia, or primary form of medicine, for 70–95 percent of people living in most developing countries. In other words, at least 4 billion people are dependent on plants for medicine, and the key ingredients in their medicine chests are getting more and more difficult to find.

This work isn't just about drug discovery; it's about developing a better understanding of the safety and efficacy of herbs that billions of people rely on for their medical care. In addition to collecting samples for laboratory research, my team also works closely with local communities and cultural organizations to help document their remaining fragments of traditional knowledge before they are lost forever.

The objective of the Nagoya Protocol on Access and Benefit-Sharing, which came into force only in 2014, is that fair and equitable sharing of the benefits should arise from the use of genetic resources. This ensures that countries share the financial benefits of discoveries linked to their genetic resources and traditional knowledge. But benefit sharing starts much earlier than the point at which a discovery might make money, which can take many years, if ever. It starts with ethically engaging with participating communities, ensuring informed consent of the individuals who contribute knowledge, and returning that knowledge to them in an accessible format. In many cases, healers not only want their knowledge preserved, they also want to know the results of the laboratory studies. They, too, want the benefits of the lens of modern science on the medicines passed down through generations in their culture.

I've found that the most successful of our international research endeavors are those I've undertaken in collaboration with scientists from the host country, and in addition to preserving and celebrating traditional knowledge of plants, we've been able to contribute to building research capacity (encompassing both infrastructure and training support) in the country through investing in student exchanges and training opportunities, joint grants, and publications. Our work in Sicily was no exception.

In all, the Isole Egadi (Egadi Islands, aka Aegadian Islands) make up a landmass of roughly thirty-seven square kilometers, and were inhabited by 4,292 people as of the 2017 census. Of the three main Egadi Islands (Marettimo, Favignana, and Levanzo), Marettimo—formerly known as Hiera, part of the Greek name for "sacred island"—is the most distant from the mainland and the wildest. The island has a rich history of beekeeping, and fragrant, pungent spiciness of wild thyme and rosemary rise from the mountainside trails, yielding a decadent honey.

Favignana is the largest of the isles, featuring a secret garden of old fruit orchards in the historic limestone quarries. Deep in the quarries, it's easy to imagine being lost in a magical world. Accessible by well-worn wooden stairs from the ground level above, it is the entrance to a network of canyons and caves where miners once hauled out heavy slabs of the calcarenite rock. In the canyons where the sunlight still kisses the ground are luscious old varieties of figs, plump with sweet, juicy flesh, and citrus trees bursting with fruit ripe for the harvest abound.

The smallest isle, Levanzo, is not without its own secrets. In addition to pastureland, it features cliffs and rocky hillsides where, in the 1600s, nearly one hundred thousand grapevines covered the landscape, though such large-scale viticulture no longer exists there today. Most noteworthy is the pastureland covered with the dead stalks of dried giant fennel (*Ferula communis*, Apiaceae). Some locals still use the large herb to make furniture because little local timber is available. In the fall when the rains come, the grounds surrounding this herbal ghost land burst forth with a bounty of king oyster mushrooms (*Pleurotus eryngii* var. *ferulae*, Pleurotaceae), considered the most delicious of all the mushrooms from the islands.

The eight of us were in Marettimo, wrapping up the last collection efforts of our expedition and hunting for a special shrub that grows on rocky slopes at higher elevations: *Daphne sericea* in the Thymelaeaceae family. This and other members of the *Daphne* genus share a rich history in traditional medicine for the treatment of everything from gonorrhea to wounds and malaria.

DURING THE COURSE of the field study, my graduate student Akram and I interviewed a group of old men, among them Alberto and Giovanni, both retired from careers balancing fishing and working their plots of family agricultural lands on the island. We sat in the shade in a small piazza by the harbor as they repaired their nets, their hands weaving in and out of the tangled web as they stitched one piece to another in a natural rhythm. We spent more than an hour laughing and talking together, discussing how things were in the old days. With their permission, Akram kept the video camera running during the conversation so that I could go back and review the tapes along with the notes that I jotted down in rapid Italian, pausing at times to interpret their thick Sicilian dialect. Akram had a knack for picking up languages, already being fluent in three. He was quickly learning many of the local plant names and some of the dialect phrases, though he was eager for our relaxed conversations in English over our evening meals. His experience reminded me of my first time in the field so many years ago, my brain both enthralled by and exhausted from all of the new bits of information it was being flooded with.

"When I was a boy," Alberto, now in his eighties, reminisced, "this island was covered with the most delicious fruits you could imagine." Other men nearby nodded in hearty agreement. He lifted his hand to the mountain, indicating the tiers of land, delineated by rocky walls, now covered in pines, and said, "None of those pines were there. Each family had a vineyard, olive groves, and fruit orchards."

"What type of fruit did people grow?" I asked.

"The largest nectarines you could imagine, among many other types of fruit," he replied, closing his eyes, smiling. "You could smell their aroma all the way from boats passing by on the sea. Now there are none," he sighed. "The pines have poisoned the land with their leaves. Nothing can grow there anymore."

This was a common thread picked up in our interviews—an example of how government policy can have long-lasting unintended consequences. After World War II and years of emigration, few were left to work the land, and the orchards began to fall into disrepair. A government-sponsored program offered to pay people what would have been a substantial fee at the time to have their lands planted with Aleppo pines (*Pinus halepensis*, Pinaceae) in an effort to reforest the region. While it proved a successful program for pine planting, it destroyed local agriculture on the islands. While tree-planting campaigns across the globe have been touted as a tool to fight climate change and deforestation, planting a single species does not a forest make. This lack of biodiversity fostered by monoculture plantings are bad for forest ecosystems and for the humans who rely on them for the acquisition of useful resources—whether food, medicine, timber, fuel, or other values.

For the Egadi islanders, their other source of income was wiped out. As a result, these communities put their efforts toward tourism, hoping that the money they bring in during the summer vacation months will last them the whole year.

"Tell me more about the wild plants," I asked the group. "Did you collect wild plants for food or medicine as boys?"

"Yes, but no longer," Alberto said as they all nodded. "We used to collect them when we went to work in our gardens. No more gardens, and many of those wild herbs are hard to find now."

"There are medicines, though, that grow near the village," Giovanni said, gesturing toward a hill near the water. "Cavallo marino grows over there."

The name translates as "sea cabbage." I assumed it must've been a wild cabbage, or in the Brassicaceae family. A tricky thing about local names though is that sometimes a single name can refer to many different species, depending on the community and region. Confirming the plant's identity and determining the scientific name is critical in ethnobotanical research.

"Can you describe it to me?" I asked. "What does it look like? Smell like?"

They said it had pale green leaves, yellow flowers, and a yellow "milk," and that it liked to grow near the sea. I was intrigued. Members of the cabbage family (Brassicaceae) do not make any "milk," or latex—and especially not a yellow one. However, plants in the Euphorbiaceae and Papaveraceae families make colored latex.

"What was cavallo marino used for?" I asked.

"When you fall or have a bad bruise"—Alberto pointed to one arm with the other, making a pained face to illustrate a fall—"you can mix it with salt, pound on it, then wrap it in a cloth and hold it on the bruise."

In the midst of our conversation, a ruckus of children laughing and dogs barking echoed down to the piazza from one of the village streets. In the center of the herd, I saw Donato and Bella (now eleven and nine years old) speaking with the village kids in Italian as they passed a soccer ball back and forth while the dogs tried to catch the ball in their mouths. The dogs belonged to everyone and no one, plump from all of the scraps they garnered.

"Mamma, guarda questo!" Donato shouted for me to watch him as he lifted the ball to deftly juggle it on his knee, eager to show off his new skill. He took after his father in looks, the same dark brown hair and olive skin. Bella was fair skinned, like me. She parted from the group and came to sit in my lap. She smelled of the wildflowers that she'd stuck into her tangled curly blond hair. "Mamma, mi dai un po' di soldi per comprare un gelato?" She sweetly asked, "Dai, per favore?" The "favore" lasted about five seconds, her hands joined in a gesture of supplication. These kids had mastered not only the language but also the bodily gesticulations of southern Italians. It was hard to say no. I pulled out a couple of euros. She hugged me in a quick thanks and ran back to her older brother, exchanging a high five as they took off in delight down a stone-paved street that led to the ice cream and patisserie shop.

While Marettimo might have been an ideal site for the research team, full of interesting plant species and ancient knowledge to explore, for the kids, it represented all the best aspects of summer. Freedom to play and run around all day with friends, unencumbered by the dangers of traffic. Marco and I could never allow them such liberties in Atlanta. We regretted that we couldn't offer our kids the same freedom to roam that we'd both enjoyed while growing up in the countryside of Italy and Florida. But here they were safe to play, and if we ever needed to locate them in the village, the network of watchful zie could always point us in the direction they'd gone. It takes a village to raise a child.

While I was out on interviews with Akram, Marco stayed at the apartment with our youngest child, a four-year-old, processing our plant collections, sorting, chopping, drying, and packing them for shipment back to the lab. Marco and I had named him Giacomo, after our favorite composer, Giacomo Puccini, and the Italian legend Giacomo Casanova. If his namesakes were any indication, this boy would be a great lover of music and beautiful women.

This wasn't Giacomo's first international trip. He was born the year I started my tenure-track job. Before he'd ever reached the six-month-old mark, his passport was covered in stamps from our trips together: Italy, where I'd returned to collect more plants; the United Arab Emirates, where I'd consulted for the national center on herbal and traditional medicine; and England, where he'd joined me for one of the annual Society for Economic Botany meetings. At the London airport, I'd been grilled by border agents; they hadn't understood why I would bring a baby to an academic conference or believed that it was the true purpose of my trip. I was exasperated by their reaction. Why was it so unbelievable that a scientist would need to bring their child along on a work trip?

Faced with the costs of after-school care for two elementary-age kids and full-time day care for our new baby, Marco and I realized that the benefits of a dual income were quickly wiped out, and we'd miss much of this incredible time in the kids' lives. We'd decided that Marco would exit the workforce until Giacomo started school—working full-time caring for the kids and Granny at home while I worked to get my career off the ground. Though it made our finances tight, it gave us a golden period of flexibility both in Atlanta and during annual summer field studies, like this one, when we needed it most. And that was priceless.

While in Marettimo, to escape the heat of the midday sun after lunch, we went on daily swims in the Mediterranean, following the rocky paths down to the shoreline. Marco and I took turns watching Giacomo play, splashing with his toy boat at the water's edge on the black-pebbled beaches. Then, with snorkel, mask, fins, and our underwater camera in hand, one of us swam along the coastline with Akram, Bella, and Donato, searching for fish and sea urchins through the crystal-clear water, keeping an eye out in hopes of spotting graceful octopuses and avoiding stinging jellyfish. I'd already had the unpleasant experience of getting stung by a jellyfish on a swim, its tentacle snaring my thigh; I shot out of the water like a bullet when it happened! Luckily, the kids, Akram, and Marco remained unscathed.

Needing to work on mainland Sicily during the week, Alfonso and Alessandro took the ferry to join us on weekends to hunt for plants and mushrooms at our field sites as we moved from island to island.

LATER THAT WEEK, working with the group of fishermen Akram and I had interviewed, we found the sea cabbage. It was the yellow horned poppy (*Glaucium flavum*), a member of the Papaveraceae (poppy) family, as I had suspected. When Alfonso, our naturalist collaborator, arrived on the weekend, he confirmed the identity. Roughly my age, in his late thirties, Alfonso is a fitness buff who has a gift for cooking the most amazing pasta dishes and decadent desserts. He's also like a walking encyclopedia of botanical names. The first days in the field with him, I'd been caught off guard as he rattled off not only the genus and species names of plants we spotted but also the subspecies and author epithets—obscure bits of intricate detail that even the best botanists don't usually file away in their minds. When I checked the names back at the apartment with my stacks of books on the Italian flora, I found that—sure enough—he'd been right every time!

I was intrigued by the local use of this poppy. It is toxic if eaten, causing brain damage and death; it's classified as a noxious weed in some parts of the United States. On the other hand, glaucine—a compound derived from this plant—is used as a cough remedy ingredient in some countries but is also abused as a recreational drug due to its hallucinogenic effects. No one had yet explored the science behind the topical therapy for bruising and pain that the fishermen had mentioned; I made a note to think about how we could investigate that in the lab.

"There are other good medicines on the island," Alberto told us. He pointed to the cluster of Roman ruins above us, midway up the mountain. "There's a lot of erva janca [white herb] by the ruins."

"Is it the good-smelling one?" I asked, familiar with this local dialect name for a fragrant tree wormwood (*Artemisia arborescens*, Asteraceae), which was also found on the other Egadi Islands and used in ritual bathing practices for newborns.

"Sì," he replied.

It was getting late, time to pack up our gear and get back to the apartment for dinner. "I'd like to climb up there to see it," I told the fishermen, "and there are some other plants I need from up higher, around five hundred meters," I continued, pointing in the distance to one of the mountain peaks beyond the ruins.

"No, no, dottoressa," Alberto said, glancing at my prosthetic leg. "It's too difficult a climb."

"Can I drive up there?" I asked, thinking perhaps someone could give me a lift.

"No," they said, shaking their heads. "Vanni has a quad bike, though; maybe he can help you. His office is in the first piazza."

I thanked them. I would check in with Vanni in the morning.

When I was eleven years old, Daddy had a good year in his land-clearing business. That Christmas, he surprised Beth and me with quad bikes. They were a pair of Yamaha Blasters that were crazy fast, and could reach up to sixty-five miles per hour at top speed! Momma gave us many lectures on keeping them in the lower gears when we raced across the cow and horse pastures or leapt over the dirt ramps Daddy made after excavating the pond. The Blaster gave me a way to play outside without need of my prosthetic leg, a huge boon especially during the weeks I had to spend in recovery from surgery or skin infections on the stump. Driving one was second nature to me, and I was hopeful that it could solve my problem of reaching the top of the mountain.

. . .

Cɪᴀᴏ, Vᴀɴɴɪ," I ꜱᴀɪᴅ.

Vanni already knew all about me—word spreads quickly in small communities—so I cut to the chase.

"I grew up using quad bikes. I'm very familiar with them. Would you be willing to rent yours to me for the day?"

"No, I'm sorry, the brakes are bad on it. Too dangerous. Plus, even if it worked, you would have to stop at the ruins; farther on the trail is too rocky."

My face fell. There had to be a way! Marco or Akram could make it up there on foot, but they wouldn't know which species to collect or what botanical clues to look for. Vanni paused, then added, "You know, you should talk to Nino, he has donkeys and mules. Maybe he can help you." Vanni directed me to the small coffee bar by the port.

Inside an open-air bar, affixed to one of the walls, was a sign that read CALL NINO FOR MULE TOURS! with his phone number listed below. Nino met me at the bar later that morning. He wore jeans and a T-shirt, his dark brown hair brushed back over his olive skin, deeply tanned from the time spent outdoors. I pulled out my map of the island and tapped my pointer finger on the mountain peak most likely to have the different species I sought.

"I need to go here, Nino."

"That's very high up," he replied, shaking his head. "The trail is rough, and I have tours in the evenings."

"We can leave early in the morning," I said, "before daybreak. Is five thirty okay? This way it won't be so hot for us *or* your animals." I gave him a big, encouraging smile.

He paused to consider it. "How many animals do you need?" he asked.

I ran a count in my head. I would definitely need one for myself the whole trip. Alessandro and Alfonso were fit enough to walk, as were Marco and Akram. I needed all hands on deck so we could work as fast as possible once up top to collect vouchers, DNA, and bulk specimens and record all of the accompanying data. It would take us at least an hour and a half each way to get up and down the mountain, not leaving us much time at the top. With all of the adults gone, that left the kids. We couldn't leave them alone all day in the village; they would need to come with us as well. Plus, Donato and Bella could help with the collections and keep an eye on Giacomo while we worked.

"How about three mules? How much would that cost?" I asked. A waiter passed by and I ordered us two espressos. Nino tossed out a number—€400, definitely a tourist price and way over what my grant budget could cover! "Nino, come on. We're scientists, not tourists, and you'll still have your animals back in time for the evening easy ride." He offered another number. I countered a bit lower and we finally came to an agreement at €250. I had to respect his bargaining strategy: just like Sahil, my business partner with PhytoTEK, had taught me, always anchor high when you begin a negotiation.

On the morning of the climb, I smeared sunscreen on the kids and me, while Marco filled our water bottles and grabbed the packs of collecting gear—clippers, plant presses, and bags. The kids grumbled at the early wakeup, but they were also excited by the prospects of the adventurous ride up the mountain. Bella, especially, had been begging me to go horseback riding for the past few months. We made our way along the village streets in the dark, carrying our packs of gear to the base of the trail situated behind Al Carrubo, a restaurant so named because of the beautiful, large carob tree (*Ceratonia siliqua*, Fabaceae) positioned at its entrance. Alessandro, Alfonso, and Akram, who were staying in a separate apartment on the other side of the village, soon joined us.

Nino was there, waiting with the mules. We lifted Bella and Giacomo to the seat of one saddle, with the plant press strapped to the pommel. Giacomo was as excited as Bella to be in the saddle, and he lovingly patted the mule's neck with his chubby little hands. Our backpacks full of collecting supplies—secateurs, collection bags, notebooks, camera, GPS— went onto the two other mules, with Donato and me each taking one. Nino had his own. All the others would walk. The first part of the trail was paved in smooth stone, placed there by the village as part of the tourist attraction to hike up and visit the ancient Roman ruins and natural spring at around 250 meters above sea level. Although paved, it was steep, and I knew that the others, despite being athletic and fit, were going to be sore by nightfall.

We made our way up slowly; I led in the front, Nino in the back. Every once in a while I turned to check on the kids. They swayed back and forth in the saddle, lulled by the peaceful rhythm of the climb. Behind them, I admired one of the most beautiful sunrises I had seen in a long time: a bright golden-orange orb crept skyward from the horizon, almost as if emerging from the sea itself.

We tied up the mules at a fountain by the Roman ruins, a cluster of military settlements (locally known as Case Romane) built in the second century BCE and an eleventh-century church likely built by Byzantine monks. I moved to collect the enticingly fragrant *Artemisia arborescens*— the "white herb" the fishermen had spoken of (aka tree wormwood)— while we had the chance. Four feet tall, clusters of the herb covered the grounds all around the ruins, dainty tufts of yellow flowers and frilly silver leaves swaying in the early morning breeze. I could understand now why this plant was so treasured as a fragrant addition to holy water for certain religious ceremonies and also to bathe newborns. Smelling it reminded me of sage, camphor, and the comforting scents of kitchen herbs; I had a strange desire to roll around on the ground with it! Steeping in a hot tub steaming with its fragrance would be amazing! Donato and Bella played hide-and-seek amid the old stones and structures of the ruins, tracing their hands along the crumbling walls and running on paths just as other children have done for two thousand years.

Near the fountain, a beautiful old azarole hawthorn, also known as the Mediterranean medlar (*Crataegus azarolus*, Rosaceae), tree grew. I went to work collecting fresh branches. I created a temporary work area on an old rock wall near the fountains. In addition to the saddled mules, we had another guest with us—one of their young fillies. She was rubbing up against me as if to cuddle, warming my heart, until I realized her true intent was to eat my freshly collected hawthorn leaf samples that I'd been preparing for the herbarium press! Nino was watching from nearby. He laughed and said, "That's one of her favorite snacks." *Sweet, naughty baby!*

After wrapping up our brief collections, we hid our bags of plants in a crook of a tree branch where they would be safe from hungry animals, with the plan to pick them up on our way back down the mountain. The broad trail quickly went from smooth, steep, and paved to a rough, narrow path with jutting rocks mid-trail. I understood then why the quad bike would have been useless for reaching the top.

As we steadily made the climb, I could see the shift in species. At the very top, the canopy of trees gave way to open landscapes covered with short shrubs—a mix of wild rockrose (*Cistus* spp., Cistaceae), mastic (*Pistacia lentiscus*, Anacardiaceae), and wild rosemary (*Rosmarinus officinalis*, Lamiaceae). We tied up the mules and brought the kids to a safe spot to sit and rest, away from the dangers of the steep cliffs that bordered the trails. Nino volunteered to keep an eye on them while I unpacked my gear and convened with the rest of the team.

Not far from where we stopped, we found abundant patches of purple flowering *Daphne sericea* atop Portella Anzine, amid the rocky peaks of the island. I had become fascinated with the *Daphne* genus after encountering its close botanical cousin, *Daphne gnidium*, during an earlier field expedition with Alessandro and Marco just off the Tunisian coast on the island of Pantelleria. Alessandro was born and raised in Palermo, where he spent his childhood hiking through the countryside mushroom hunting. He and Marco shared a similar physique and love for running and cycling. Pantelleria had been our first expedition together, and Alessandro was a natural not only in interviewing local people, but also in spotting mushrooms in the forest that I would have walked right past. Plus, he actually knew how to identify them—if not by eye on the spot, then with the aid of a microscope back in his lab at the University of Palermo.

On Pantelleria, we'd learned that *Daphne gnidium* was used for everything from a pest repellent in dog pens to a hemostatic agent for humans. It also has compounds with antiviral activity. I was eager to bring its botanical sister (*D. sericea*) back to the lab to examine the chemistry and bioactivity of this relative to see whether it had any antibacterial activities as well.

Besides this *Daphne* species, we also looked for more populations of the abundant mastic shrubs and Sicilian sumac (*Rhus coriaria*), which share a long history of treating gingival inflammations and oral infections. Both of these species are members of the Anacardiaceae family—a group that includes poison ivy, cashews, and mangoes, but also the medicinal plant *Schinus terebinthifolia*, or Brazilian peppertree, which we studied in the lab.

Just 770 miles due east as the crow flies is the Greek island of Chios, situated in the Aegean Sea. The island is famous for the production of "Chios tears," which is actually the dried resin tapped from the same species of mastic I was sampling from in Marettimo. Mastic resin is one of the plants that appear in the Bible, and it has been used as a chewing gum for at least two millennia. It has also served as an important medicinal spice ingredient throughout history and was used to treat gastrointestinal complaints. Though some scientists had examined its antibacterial activity in lab settings, much work remained to be done regarding evaluation of the full spectrum of bacteria it could be effective against, as well as which chemicals in the plant were most responsible for these effects. Similarly, Sicilian sumac has been used as a spice and medicine since at least the time when the region was ruled by the ancient Romans. Like mastic, it required further study. I added them both to our growing collection, tying bags full of the leaves and stems to the saddle of one of the mules.

Marco took off down a narrow, rocky path that was too rough for me: he knew which oak (*Quercus*) species to look for in that area. He had always been a plant guy, exploring his family's land as a kid. But just as he taught me how to really cook, I taught him to really plant hunt. Akram and I focused on sorting through the shrubs that flourished in the open sunny fields at this elevation. Each collection required careful documentation of the plant's characteristics, habit (herb, shrub, or tree), and habitat (what the landscape looked like) to accompany the herbarium voucher collections. We worked quickly, but the hours flew by. Soon enough, we had to pack everything up, load it onto the mules, and start the journey back down the trail.

Upon our return to the village, hours of additional work awaited. We still needed to separate the various plant tissues and load them into our field dryer. We also had specimens from previous collections to grind up and vacuum pack. Donato and Bella took off to enjoy the rest of the afternoon playing soccer with their village friends. Giacomo was out for the count with a nap—the morning had been long and tiring for us all.

Later that week, we wrapped up the expedition by presenting our preliminary findings to the village in a workshop co-organized with the local cultural organization and history museum. This is one of the ways that I give back to communities, by ensuring that the information we collect on traditional uses of plants and fungi is documented and shared in an accessible way. We've done similar things with our work in different locales. In the Monte Vulture area of Italy, with the support of EU funds, Andrea and I worked with community leaders to establish a botanical garden for use by the local schools. I wrote and printed more than one thousand copies of a bilingual book on the traditional knowledge of wild foods, medicinal plants, and healing rituals that were freely distributed to villages throughout the region.

I planned to return to the Egadi Islands later that fall to capture the peak season for wild mushrooms. Working with Alessandro and fellow ethnobotanists from the United States and the UK, Dr. Maria Fadiman and Susanne Masters, we were successful in collecting the remaining species that had been mentioned in our summertime field interviews but that hadn't been available in the wild at the time. Sometimes return visits have to be planned around the growth and harvest season for the targeted species; nature works on its own schedule.

BACK ON CAMPUS IN ATLANTA, I stared out into the surprisingly crowded lecture hall. I waited for the students to take their seats, the latecomers shuffling past the early birds, trying not to tip over in the amphitheater-style seating. The courses I developed for the Center for the Study of Human Health began with a meager enrollment of fourteen to twenty students, but now, six years later, my introductory Food, Health, and Society course regularly drew some one hundred students, and my more advanced senior-level course, Botanical Medicine and Health, had around sixty students each year, much higher than the typical elective class of twenty-five. I'd worked diligently to develop and enrich the coursework with lots of hands-on demos in the lecture hall to help the students discover more ways to connect with the plants that influenced their health. In 2019, Emory recognized my efforts with a distinguished undergraduate teaching award. I also took my passion for teaching to the broader public that year, working with Rob Cohen and Christine Roth, a team of Hollywood producers I met through my consulting work, who encouraged me to share my lessons on food and health on a new podcast series. With their help on production, I began taping interviews with experts and sharing my knowledge of plants and the food-medicine continuum; the *Foodie Pharmacology* podcast was born.

On this particular day, as the students finally stopped fidgeting, I drew a deep breath and began my lecture on one of my favorite topics: oral health and hygiene.

"Good morning," I announced, clipping on my lapel mic. "I want you all to think back to a few hours ago when you woke up today. Did any of you run your tongue over the surface of your teeth this morning?" Several of them nodded in affirmation. "Can you describe what it felt like?"

Amy, in the front row, had her hand up. "It felt kind of rough, kind of gritty."

"And does anyone know what causes that gritty texture on your smooth teeth?"

Craig raised his hand and I nodded to him. "Bacteria!" he responded.

"That's right. Can you also recall from our lecture on infectious disease how these bacteria are stuck on your teeth?" I continued.

"Yeah, that's a biofilm, I guess," he responded.

"Correct! Just as microbes find ways to stick to surfaces in nature, such as a slime-covered rock in a stream, the same thing happens in our own mouths. That gritty feeling you sensed is a community of many different oral species that have come together to create their own special city right there on your teeth. In fact, there are an estimated five hundred

different species that live in each of your mouths!" I went on to describe
how the mechanical action of a toothbrush with the abrasive ingredients
found in toothpaste—such as hydrated silica—can act together to remove the dental biofilm biomass that forms on the teeth.

I opened up a large cardboard box at the front of the lecture hall filled
with individual twigs sealed in blue plastic wrappers with the word *Miswak*
stamped across the top. My graduate students Akram and Lewis were
working as my teaching assistants and distributed them to every student.

"For thousands of years before we ever had toothbrushes, mint-
flavored mouth rinses, modern chewing gums, and cinnamon-spiced
breath mints, people relied on nature's resources for their oral health and
hygiene. Here's an example of a traditional chewing stick. Miswak comes
from the plant *Salvadora persica* in the Salvadoraceae family. Think of it as
a natural toothbrush that you can simply clip back with each use and
toss out when you're done—it's 100 percent biodegradable."

Akram and Lewis demonstrated how to trim back the bark to expose
the frayed brush-like fibers of the twig and use it as a toothbrush as I
continued: "Miswak is not the only chewing stick used in oral hygiene.
There are many others. One other you may recognize is that of the neem
tree, or *Azadirachta indica* in the Meliaceae family. While these two plants
are not closely related, they do share common qualities of exhibiting
antimicrobial properties useful for keeping some of the microbiota that
are detrimental to oral health in check."

As students tried the chewing stick, some made faces of disgust while
others had faces of delight. I noted that unlike our plastic toothbrushes,
which can have biofilm buildup on them as well if not frequently re-
placed, chewing sticks were actually cleaner: after chewing the stick each
day, the user simply cuts off the tip and new bristles are exposed until the
twig becomes too short for ease of use. Then a new twig can be harvested from the plant, which can be grown outdoors. Plus, no water or

toothpaste is necessary for this practice, making it a very convenient travel toothbrush I take advantage of myself when on hiking and camping trips.

The lesson continued with a discussion of teeth-blackening practices of Indigenous tribes in Southeast Asia and Polynesia, in which the teeth are painted black with a solution made of ferric acetate (from iron) combined with plant tannins. Plant galls, or tissues that develop in response to pathogen or pest attack, much like a tumor, are especially rich in tannins and are frequently used as an ingredient for this process. In addition to producing a mouthful of ebony-colored teeth, this practice protects the teeth from decay instigated by microbial biofilms. Tannins, like those that my research team had found in blackberry roots and white oak galls, were exceptionally effective in blocking biofilm formation by bacteria.

Oral health and hygiene practices using plants remain a fascinating, and yet heavily understudied, aspect to research on the medicinal potential of plants. There were so many things that scientists could do with these plants if we only had a deeper understanding of their efficacy, utility, and means of sustainable production. I told the class, "Imagine the reduction in plastic waste alone that we could have across the world in terms of toothbrushes, toothpaste, and all of the other petroleum products that go into oral care if people grew their own botanical chewing sticks."

After class, I headed over to the Micro lab to check in with my team. Over the past few years, my lab had continued to grow. More than a hundred undergrads had trained in either the lab or the herbarium—in many cases, both. One of my undergraduate honors students, Danielle, was there in the lab, in the process of placing microtiter plates full of plant-extract-treated *Porphyromonas gingivalis*—the rod-shaped bacteria responsible for periodontitis—into the anaerobic growth chamber.

Poor periodontal health from *P. gingivalis* infection has been associated with numerous health conditions, ranging from osteoporosis, cardiovascular disease, diabetes, rheumatoid arthritis, obesity, respiratory infection, and preterm birth. Emerging research has also linked this pathogen to inflammation of the nervous system, with some studies showing possible links to neurodegenerative diseases including Alzheimer's, cognitive decline, and dementia.

Danielle shared her latest findings with me. Out of thirty plants that she had cherry-picked based on ethnobotanical use reports for oral health and hygiene, she'd identified 109 extracts in the Quave Natural Products Library (QNPL), my chemical library of plant extracts and isolated plant compounds, to test against *Porphyromonas gingivalis*. Eleven proved efficacious, with an extract of the mastic fruits I'd collected on the island of Marettimo performing better than all of the others.

"Keep it up," I said. "I'm going to check in with Micah on last night's run with the new extract fractions."

Micah's undergraduate honors research had examined medicinal plants of the southeast United States used during the American Civil War, and after graduating with highest honors, Micah returned to my lab to work as a research specialist for an additional two years before starting graduate school in agronomy at the University of Florida.

Micah sought to isolate compounds with antibiotic activity against multidrug-resistant *Acinetobacter baumannii* infections. This pathogen causes serious infections in the blood, brain, lungs, and wounds. There was a surge in antibiotic-resistant infections—especially debilitating soft-tissue infections—among US service members injured in the Kuwait and Iraq region during Operation Iraqi Freedom and in Afghanistan during Operation Enduring Freedom, and as a nosocomial (hospital-acquired) infection. *Acinetobacter* is exceptionally adept at surviving in the hospital environment: it can linger on dry surfaces for up to thirteen days. That's *ten more days* than most other gram-negative pathogens! So a patient being treated in the same room nearly two weeks after someone infected with *A. baumannii* is moved is at risk of infection should any contaminated surface (such as a bed rail) not be extensively sanitized. There's a reason my team keeps multiple spray bottles of bleach or 70 percent ethanol on hand in every room of the Micro lab. We clean our work areas with religious zealotry.

Working both on our Brazilian peppertree extracts from Florida and some Sicilian sumac that we collected in the Egadi Islands—both members of the Anacardiaceae family—we had noticed strikingly similar properties between these extracts in our tests against both a highly drug-resistant type of gram-negative bacterium known as CRAB (carbapenem-resistant *Acinetobacter baumannii*) and an emerging fungal pathogen that exhibited multidrug resistance and high fatality rates in patients (*Candida auris*). CRAB hospitalized 8,500 people in the United States in 2017, killing 700, while the super-resistant fungus *Candida auris* popped up in the United States in 2015, and cases have been spreading ever since. In addition to being incredibly difficult to treat—and in some cases, impossible—both pathogens linger on surfaces and medical equipment, setting up unwitting patients for potential exposure to deadly infection while they undergo treatment for other conditions. Our sumac and peppertree extracts exhibited growth-inhibitory action against even the toughest of these strains, and working with collaborators Dr. Julia Kubanek (a natural products chemist at Georgia Tech) and Dr. Dan Zurawski (a microbiologist at the Walter Reed Army Institute of Research), we were making real headway.

Our first breakthrough was the successful isolation and identification of a major active compound they both shared: pentagalloyl glucose, or PGG. Importantly, even after dosing the CRAB with low-dose treatments for three weeks—which usually makes any other bacteria resistant to the drug through the rise of mutants—we didn't see any resistance with PGG! Now, PGG has downsides—it may not be optimal for systemic delivery in the body—but these issues may be tackled by teams of medicinal chemists and drug formulators. We could also explore its utility in topical treatments of infection, perhaps as a medicated bandage, rinse, hydrogel, or ointment.

To do so, though, first we needed to obtain more grant funding. The pattern of finding leads—extracts that worked in our models of infection—and taking them forward, closer to the clinic, had become a part of the research group's rhythm and workflow. The initial finds from our screening efforts were followed by work to isolate and identify the different compounds involved; sometimes there were multiple compounds required to achieve the greatest efficacy in the fight against an infectious threat, such as in the case of the 220D-F2 blackberry extract. Then came the mechanistic studies to identify how these compounds worked and if they were safe to human cells. This was followed by animal models of infection led by our collaborators who are experts in those specific pathogens. As each month passed and our tool kit grew along with the experience of our team members, our rates of success grew, too.

Although I'd been busy in building up my academic lab, PhytoTEK and our patented anti-biofilm technology from the blackberry extract were never far from my mind. We had joined forces with Firefly Innovations, a Boston-based biotech company focused on the development of medical devices for the wound-care space. Every year, 6.5 million people suffer from chronic wounds in the United States, and this places a $28 billion burden on the healthcare system! Firefly proposed to integrate our blackberry formula in their eco-friendly and all-natural antimicrobial bandages for dressing chronic wounds, where biofilm infections flourished. This was an ideal arrangement. My expertise is in the discovery side of things, but it takes an entirely different and highly nuanced skill set to develop a drug or medical device and successfully shepherd it through the trials of FDA approval. I was glad to have partners in the game that I could pass the ball to in our rush to the end zone.

IN THE EVENINGS AT home, our family dining room table buzzed with chatter. No matter how busy things were, I made sure to be home for dinner. While Marco shuttled Donato and Bella back and forth to soccer practices, I'd pick up wherever he left off on prepping dinner as Giacomo played with his toy cars in the living room. Mealtimes were treasured opportunities for us to all reconnect at the end of our lived experience each days. After serving the food, we went around the table taking turns as we shared the exciting moments of our day. Granny, now in her early nineties, regaled us with her theories on alien invasions and how the pyramids were really built—all "knowledge" gleaned from her binge-watching of conspiracy TV shows in her room. Bella and Donato shared spirited stories about shots and saves from soccer practice and their mastery of multiplication tables or social studies. And Giacomo showed off drawings he'd colored with Marco. After finishing his dinner, he crawled into my lap and made his nightly plea for me to tell him a story. He didn't want me to *read* a children's book but rather invent a story in which he was the protagonist. His favorites involved fire trucks and the rescue of a kitty from a tree. I'd start the story, "Once upon a time, there was a little boy named . . . ," and before I could finish the line, he'd eagerly jump in: "MoMo! Mamma, the boy's name is MoMo." MoMo was a nickname of his own creation. He hadn't been able to pronounce his name early on, calling himself Jack-ah-moh-moh. So he'd shortened it, and the name stuck.

CHAPTER 10

Billy Fell off the Swing

"Who in the world am I?" Ah, that's the great puzzle!

LEWIS CARROLL, *ALICE'S ADVENTURES IN WONDERLAND*, 1865

Tuberculosis (TB) is well established as a global killer—its slow-growing lifestyle and propensity to acquire resistance present major hurdles to successful treatment. My maternal great-grandmother died of tuberculosis in the pre-antibiotic era of the 1920s, leaving behind her husband and young daughter, my grandmother. She was one of many who succumbed to the disease as she withered away in a sanatorium. Today, while largely absent in the developed world, TB still kills 1.4 million people across the globe every year. We need new drugs to fight it.

And so I was out in the field in south Florida with Kate, my gifted lab tech and fellow Floridian, and Ferris Jabr, an experienced journalist who'd flown out from Portland, Oregon, to join us on the hunt. It was his first day in the field with us.

"I'll keep an eye out for gators and water moccasins while you wade in the creek," I told Kate.

While I appreciated the beautiful clusters of yellow water lilies (*Nuphar lutea*, Nymphaeceae)—broad-floating leaves and lemon-yellow flowers above the tannin-stained water—my interest, as ever, was driven by the plant's traditional uses and appearance in folklore. Historic records detailed its applications for everything from food to witchcraft, plus the myriad medicinal applications including for pain relief and treating TB, gonorrhea, hemorrhages, fever, heart disease, and skin boils. The boils, gonorrhea, and TB are what captured my interest as a clue that it might yield some interesting antibiotics.

I wished I could've joined Kate and Ferris in the water, but I couldn't get in the creek without damaging my prosthesis. Buying a new leg is like purchasing a car. I'd destroyed legs in the past through exposure to salt water and muck that rusted out the screws and jammed the lock-pin mechanism that held my stump in the socket. Insurance covered only so much of the costs in building a new leg: I was allowed one only every three or four years. I had another year to go before I could request a new one.

I'd been hesitant to allow Ferris on the trip. I'd communicated with journalists plenty of times before—over-the-phone interviews about my team's latest scientific papers and discoveries that had made headlines— but I'd never allowed a journalist to accompany me in the lab or in the field. My time in the field was precious and limited; I couldn't afford distractions for myself or my team. He'd been persistent, though, and after reading some of his articles, I found that I admired his writing style. He wrote about biology and ecology like a poet, his words flowing and yet also crisp and clear. I relented, but under one condition—if he came, he'd also have to pitch in and work with the rest of the team in collecting and processing plants.

Ferris maneuvered the shovel into the thick mud, attempting to loosen the roots of a lily while Kate tugged at the top of the plant.

"You've almost got it!" I encouraged them. Then, with a final yank, they lifted the plant from the water's surface. "Oh my gosh! It looks like you just harvested the predator head from *Aliens*!" I doubled over laughing, and Kate and Ferris joined in, tossing the plant up onto the sandy creek bank next to where I stood, its roots dangling like long dreadlocks.

"Come on, let's get you dried off," I told them after we finished collecting, bagging, and field pressing the sample to be dried back at our base camp.

Sitting in folding chairs in the shade of an old oak tree while snipping our samples for drying, Ferris took out his audio recorder to interview me. He explained that he'd pitched the piece to *The New York Times Magazine*, and that it would likely appear in print later in the fall. "That's cool," I said. I'd obviously heard of *The New York Times*, but I hadn't ever seen the magazine before. It didn't seem like a big deal to me. Mostly, I was excited about having the extra set of hands on board to help with collecting during the expedition.

AFTER WRAPPING UP OUR spring collection trip in south Florida, I sat in my usual spot in the Phytochem lab, adding more than one hundred species to my growing collection of wild medicinal species for chemical extraction and lab testing. Each species was separated by tissue type and then either extracted in 80 percent aqueous ethanol or boiled in water, before being dried down for long-term freezer storage. In the Micro lab, students loaded these new extracts onto "master plates"—ninety-six-well plastic dishes that would be filled with specific concentrations of the plant extracts for testing against our panel of superbugs.

Our current campaign included a search for any extracts that could inhibit the growth of the multidrug-resistant ESKAPE pathogens—an acronym used to describe bacteria that often "escaped" antibiotic activity. This included high-priority bacteria, for which new antibiotics are desperately needed: *Enterococcus faecium*, *Staphylococcus aureus*, *Klebsiella* *pneumoniae, **A**cinetobacter baumannii, **P**seudomonas aeruginosa,* and ***E**nterobacter* species. These were deadly killers. Take *Pseudomonas aeruginosa.* In 2017, drug-resistant strains of *P. aeruginosa* hospitalized nearly 33,000 patients in the United States, killing 2,700. It's particularly dangerous to children with cystic fibrosis (CF), an incurable inherited disorder that causes damage in the lungs and digestive system. More than 70,000 people across the globe live with CF, with the majority of cases diagnosed by the age of two. The disease affects the lung cells that produce mucus, creating thick and sticky fluid, and the lungs of these kids are first afflicted with staph bacteria, and then *Pseudomonas* sweeps in. CF patients require frequent antibiotic treatment to battle the lung infections, and antibiotic resistance is a major barrier to their survival.

To do this work, I needed help. After a spate of undergraduate recruiting over the past couple of years to meet the ambitious timelines I'd set on projects in both the labs and the herbarium, I'd earned a reputation as a strong and supportive mentor. Now, every semester I went through the process of reviewing twenty applicants for research internships, interviewing ones that passed preselection, and then delivering the good news to the select few that landed a slot in the lab, and the bad news to the ones I didn't have room for. I was tough in the interview process and made my expectations very clear.

While the typical lab might have a postdoc, a couple of graduate students, and maybe one or two undergraduate interns, I needed more skilled hands to meet my goals. I supervised research for ten undergrads in the labs and another ten in the herbarium each year.

"If you're selected to join the research group," I told the applicants, either college freshmen or sophomores, "I expect a commitment from you to continue in research over the next three years—for twelve to fifteen hours per week during the school year. If you work on an independent project in the summers, it's a minimum of twenty hours per week.

You'll work at your own pace. After completing the online training modules we've developed, passing the lab protocol quizzes, and demonstrating your skills hands-on to one of the senior lab members, only then do you get to advance to the next skill level in experiments.

"If you start off in the herbarium, you'll be on the top of the wait list to get into the lab in the next cycle if you do well there. My best students have successfully trained in the herbarium and in the Phytochem and Micro labs, balancing out their skills across the full spectrum of natural products discovery from plants."

I was asking a lot. I'd once been in their shoes, though, in high school as a volunteer in the ER. In college, I'd taken on a number of work-study lab jobs—first in a cancer research lab, next in a neurobiology lab. In the cancer lab, I was tasked with weighing out amino acids on a scale—it was monotonous and redundant. In the neurobiology lab, I cared for a colony of grasshoppers kept in large cages in a walk-in incubator. Before entering the steamy hot room, I had to change into a hazmat suit with a mask to protect my lungs from their shed particles that filled the air. After cleaning out the tray of waste from beneath the cages and swapping out new heads of romaine lettuce for the half-eaten remnants, I had to pull out any dead insect bodies. Unfortunately, anytime that a wily one was able to escape the cage during this process, it had to be captured and destroyed. People walking down the hall could glance in to see me waving the arm of a shop vacuum through the air as I chased down the errant escapees, sweat running down my face and inside my mask. It took me years to be able to eat romaine lettuce again without gagging from the memories.

No one in either of those labs ever even explained what the purpose of these exercises was or what the main research questions being explored were. I was never invited to lab meetings or encouraged to read research papers from the lab team. I lasted only a semester in each position before I quit. I wasn't the quitting type, but these jobs were without a sense of purpose.

I was determined to do things differently in my lab. Every member would be informed about the rationale behind each task and its importance to the overall research mission. Even if it was an arduous task—as our sweaty, exhausting field expeditions were—at least they'd reap satisfaction from knowing their efforts made a difference, ultimately knowing that their contributions counted toward the advancement of science and medicine. I'd learned the lesson of how a toxic work culture infects an entire team during my early days of training. I made it clear that everyone in the group was to be treated with respect and held to high expectations, regardless of rank or number of years in training. I had a zero-tolerance policy for bullies and harassers.

These were high expectations for an unpaid job, but these research interns learned from others that if they got onto the team and proved themselves, I'd grant them the gift of training and an unusual degree of creative freedom. I wasn't interested in students looking for a résumé booster with a semester of lab work. If they needed money, I made sure that they got it, securing work-study positions for them. If they needed peer-support networks, they readily found it in the team. I emphasized unity and teamwork, reinforcing it with Quave Research Group and Herbarium T-shirts that the students themselves designed. Outside of the lab, we enjoyed group hikes, trips to the botanical gardens, picnics, camping and botanizing trips, summer barbecues, and holiday parties together. Regardless of rank as a younger student or a senior PhD scientist, all were part of a team. Armed with strong research experience and glowing references, plus a track record of publications and poster presentations, many of these undergrads went on to top graduate programs to successfully pursue their MD, PhD, PharmD, MPH, or a combination thereof!

As more funds came in with grants and fellowships, I secured more slots to train new graduate students from the pharmacology and microbiology programs. We were also joined by more and more scholars from all over the world: Africa, Europe, Asia, the Middle East, the Caribbean, and South America. With seven million subscribers to *The New York Times Magazine* across the globe, Ferris's feature article on my work *was* actually a big deal, after all. It was followed by features in other international outlets, such as *BBC Focus*, the Mexican magazine *Valor*, and the German beauty and fashion magazine *Brigitte*! Our research was covered in the children's science magazine *Ask*, and I was invited onto radio shows on NPR and a slew of podcasts. These features gave us tremendous exposure and offered a unique platform to highlight not only the importance of our lab research, but also the value of the Emory Herbarium as a natural history collection and research resource.

Collections of plant specimens are fundamental to our understanding of global plant diversity. Today, nearly 2,000 new plants are being named each year—each year!—and yet 2 in 5 plants are estimated to be threatened with extinction. Botanists are in a race to identify and track these species before they're lost forever. Almost 450 years of collecting and cataloging plants has resulted in the need to curate more than 392 million specimens in 3,324 active herbaria spread across the globe. Yet, over the past 50 years or more, herbaria have suffered increased neglect and budget cuts. As biology programs move away from natural history and botany in favor of technological-driven molecular approaches to science, we've trained fewer botanists, and the curators of these centuries-old scientific collections have been left to scrounge for the most basic operational support. This wasn't just an Emory problem.

Our donor funds for the herbarium were running out and I was beginning to panic. Just when things were coming together scientifically, budgetary hurdles could bring it all crashing down. So, to highlight the importance of the herbarium, I ensured that photographers for these magazine pieces had access to it, and soon exquisite photos of a collection of its specimens graced the pages of *National Geographic*! This was followed by video clips aired on our local NPR station's website and TV documentaries on PBS in the United States and networks in Europe. The attention to the collection bought us a temporary reprieve with some funding from the university while I continued to look for a long-term solution. Because I took the time to discuss my research breakthroughs with journalists and engaged in science communication for the general public, our global reach to other scientists had also expanded. People across the world were learning about our papers in *The Telegraph* and *The Washington Post* and on NBC News and CBS News, plus a myriad of non-English international news outlets in Polish, Italian, Spanish, Mandarin, Portuguese, French, Arabic, and more. I began receiving streams of emails from scientists in other countries who wanted to come train with my team, if I was willing and had space to host them. There were students, postdocs, and even professors who joined us in the lab for periods ranging from two weeks up to two years, their expenses paid for by prestigious grants they'd earned on their own through Fulbright fellowships or competitions for research funds specific to their home countries. At one time, we counted the number of languages spoken fluently in my group and reached a total of eight!

Upon return to their home countries, many of the visiting scholars would become long-lasting collaborators with the team, expanding our research network and access to diverse plants and ecosystems across the globe. My vision for the growth of our research enterprise was taking shape, person by person, and discovery by discovery.

Still, I faced strong headwinds—and not simply from lack of funding or other scientists remaining skeptical of our ethnobotanical approach to drug discovery. You don't have to be an ethnobotanist to battle these adverse conditions.

Aʀᴇ ʏᴏᴜ Tᴏᴍ's ᴅᴀᴜɢʜᴛᴇʀ?" a professor in his sixties asked me, while peering through his bifocal lenses. About half his age, I sipped a glass of white wine and looked at him. Let's call him Professor Toad.

By "Tom's daughter," Professor Toad was asking if I was the offspring of Tom, host of this little faculty-recruiting party and a department chair at Emory. I was there to chat with and hopefully encourage the potential recruits to accept an offer and join our university.

I wasn't a member of Tom's department, but I understood why I had been invited. One of the potential recruits was a super-talented female scientist, and the department wanted to show her examples of female faculty working with other members of the department. For the most part, these events are enjoyable times of catching up with colleagues. The problem for me is that they also take away from time with my family, and time had become my most precious commodity. So I did not accept the invitation lightly.

This particular evening came on the heels of a week of exciting press on my research group's latest publication on the virulence-blocking activity of peppertree compounds. The paper caught the eye of several international news outlets, and my face was plastered across the university website's home page for almost an entire week. The daily communications emails to the university faculty and staff included links to all of the articles, and my picture was featured on mass emails with the top news of each day.

"No," I replied to Professor Toad. "I'm in the dermatology department."

I was pretty sure that Tom, our generous host, didn't have a daughter.

"Oh. Where's your husband? Maybe I know him," he said.

My amazing and supportive husband was at home watching our three young children—taking care of dinner, bath time, and bedtime while I was at this work event.

"I don't think you know my husband. *I'm* faculty in dermatology."

"But you look so young!" he sputtered.

I was grateful we'd cleared everything up.

"But so," he fearlessly went on, "who do you work for?"

"Work for? I'm a PI [principal investigator]. I run my own group of thirty students, postdocs, and visiting scholars. I work on natural products drug discovery from plants."

As if the conversation could get any more tedious, Professor Toad then began to explain to me what natural products chemistry was, and with many inaccuracies. He insisted that he connect me with an expert he knows who could "help" me.

"Thank you," I said with a deep southern accent, my best impersonation of someone who is grateful. I held back from adding, "Bless your heart." Ask any southerner about this phrase; said on sugar-sweet lips, it is not necessarily a blessing or an expression of sympathy, but rather a polite way of masking an insult, the crux being "Wow, you are *really* stupid or impaired, but *bless your heart*, you just can't help it."

Instead, I caught myself, took a calming breath, and said, "But I am an expert in this field. Why don't you google me? Actually, why don't you do it right now?" I gestured to the smartphone in his hand. "Here's how you spell my name."

Professor Toad scrolled through his phone, browsing the link titles, and then looked back at me with a sheepish expression. It appeared that Professor Toad held a dual PhD in immunology and mansplaining.

He wasn't the first man to behave like this. The questions are all as familiar as they are enraging: *Who do you work for? But you're too young to be a professor. Whose lab are you in? Oh, I thought you were a graduate student. Where's your husband? You travel so much, won't your husband be upset? Where are your children? Who takes care of your children? Shouldn't* you *be with them?* I've dealt with bullies at every stage of my life, and the trend shows no signs of letting up. In grade school it was *She only won that award, you know, because she has a fake leg.* Today it is *They only asked her to be on that committee because they needed a woman for diversity.* It's hard to take the higher ground, especially when you are systemically pushed down. When the words and actions of those around you repeat over and again that you don't belong, it's almost impossible not to at least entertain the idea.

After spending long days in grant study section meetings, usually with a ratio of twenty male to one or two other female reviewers, I've been cornered in hotel elevators by some of those same men looking for a sexual liaison. This after displaying the utmost professionalism and expertise while debating the merits of scientific grant proposals. To have to extricate myself from the situation firmly but carefully to avoid creating an enemy in my field that could tank my future grants is tiring, annoying, and, frankly, scary. The next day, I sit across the room from them in meetings.

Take Professor Creeper. After enjoying a lovely morning of meeting with students and faculty and then delivering a talk highlighting my lab's latest exciting research at a seminar for another university, I was approached by an older male professor. He walked up and got incredibly close to me. I felt his stale breath, smelling of coffee and cigarettes, blow against my face.

"It's so nice to have a speaker that was so *beautiful* and intelligent here."

I took a step backward. He took one closer. His breath felt hot on my cheek. He was inappropriately close, leering at me, making me intensely uncomfortable. This was happening in a crowded room. No one seemed to notice. I'd never met Professor Creeper and had no prior professional interactions with him. Should I have walked away? Shouted? In retrospect, maybe I should have called him out right there. But the entire encounter took me off guard and I was frozen in shock, not knowing what to do. When a student approached me for a follow-up discussion of the talk, Professor Creeper finally left me alone.

That night as I turned over the events of the day in my head, I worried about how Professor Creeper's behavior affected women at that university, especially students, postdocs, and young faculty. What did this do to basically anyone who fell below him in the hierarchy of academic power in that department? I was surely not the first. I emailed the dean of the school in the hope that they would look into Professor Creeper's actions. The response was a stack of paperwork for me to file and vague assurances that they would investigate.

Sexual harassment is an enduring stain on the fabric of science.

This is not new in science or other disciplines. Some of the most well-known names in my field have behaved horrifically toward colleagues. In his account of the discovery of DNA in *The Double Helix* (1968), Nobel laureate James Watson wrote of Rosalind Franklin—a brilliant chemist who made critical contributions to the discovery of DNA and who many thought should have also been awarded a Nobel Prize for her efforts—in such sexist terms that it disgusts and pains me to imagine what her workday encounters with him and other contemporaries must have been like. "By choice she did not emphasize her feminine qualities," Watson wrote. "Though her features were strong, she was not unattractive and might have been quite stunning had she even taken a mild interest in clothes. This she did not. There was never lipstick to contrast with her straight black hair, while at the age of thirty-one her dresses showed all the imagination of English blue-stocking adolescents. So it was quite easy to imagine her the product of an unsatisfied mother who unduly stressed the desirability of professional careers that could save bright girls from marriages to dull men." Also, Watson, this woman's work was key to discovering the double helix! The clear belief that her intellectual contributions meant nothing, that, in fact, she existed solely for the pleasure and entertainment of the men around her—*and that she failed at it!*—is deeply unsettling.

I first read *The Double Helix* as a college student for one of my biology courses. I felt utterly depressed to see this in print from a man whom my college professors seemed to revere. Watson went on to engage in a scientific career full of accolades and awards, securing prized leadership roles such as the director of Cold Spring Harbor Laboratory and involvement in establishing the Human Genome Project. He also went on to continue in sexist and, later, racist rhetoric, asserting that certain races (especially dark-skinned individuals) had a stronger sex drive and that there were differences in IQ between races due to genetics.

And yet science promotes such ridiculous demagogues into positions of power. They are venerated, empowered, and even idolized without much consideration of the less savory parts of their beliefs and behavior. Men like Watson are often excused as "products of their time." What impact does this have on future generations of talented students curious about science? Or to those of us struggling to maintain a foothold in the ivory tower? Indeed, what are we to make of the impunity given to those in power who use their platforms to do harm?

Success in science today is largely driven by money, and where there is money at play, nepotism and the old boys' network are never far away. Buddies help out buddies—whether it's by being a vocal advocate on a grant review panel to ensure a better funding score or making connections that propel one another into the spotlight. More often than not, those buddies helping one another along are men who leverage their privilege to climb even higher in the ranks. After a while, this success, like a colony of bacteria, is self-propagating. In other words, the more grants or honors one has, the lower the bar to clear in getting the next grant or honorific award.

Although women now represent a large portion of the graduate student body in many STEM fields—receiving about half of all doctoral degrees—there is a bottleneck at the key transition period from trainee to assistant professor, and an even tighter bottleneck in the tenure-track promotion path to full professor. While women are being trained in the sciences at higher rates than ever before, we are hugely underrepresented in senior positions in our respective fields. Part of this has to do with timing: due to the lengthy years of training, around ten years of graduate and postdoctoral training, the point at which women enter the workforce in their thirties often coincides with the time they may choose to start a family.

I was unusual among my peers at FIU for starting to have children in grad school, but had I waited for what some would consider the "right time," that is, having secured tenure and thus job stability, I likely would have lost my chance. Marco and I wanted a big family, and my constellation of complicated medical issues made each year I waited even more dangerous to both my health and any potential children. Plus, though I might have been an outlier among my graduate-school peers, it was a normal thing to have kids in your early twenties where I grew up.

The choice of motherhood during graduate school and then again in the earliest days of my career came with its own costs beyond the day-to-day struggles or finding work-life balance, and I'm not alone in this experience. A global survey of women in academia found that women having children soon after their PhD are less likely to get tenure. At the time of writing this book in 2021, I'll have been in my tenure-track job for eight years and promoted to the level of associate professor, but to my great despair, still without tenure.

Moreover, an economic study of gender differences in early career STEM professionals found that women, and especially mothers at this stage in their career, suffer salary penalties. I've seen this firsthand with my female colleagues in academia—especially in private university settings where professors' salaries are not made publicly available. Oftentimes, women only learn of salary disparities with men at a similar rank, accomplishment, and career stage when they put together joint grant applications with colleagues: the difference could be as low as a $5,000 annual salary gap, but in many other cases it is $40,000 to $50,000 or more—more than Marco made in salary in a year. This annual amount adds up over time—funds that could've gone toward the purchase of a home, or retirement, or childcare costs. Year after year, the chasm widens, and when trapped at a lower starting tier, no matter how many performance-based raises they earn, women can never catch up.

Another study found that 42 percent of mothers and 15 percent of fathers leave full-time work in a STEM field soon after having children. The causes behind this phenomenon are complex, and include factors ranging from gender discrimination to greater caregiving and homemaking responsibilities for mothers in particular. Unlike working fathers, working mothers are faced with societal stereotypes, such as being painted as too focused on their children and thus less reliable for work. The maternal wall to career advancement and stability in science is real.

Every evening networking event or work trip came with a cost-benefit analysis that Marco and I considered together. We were a team, and in my absence, he'd carry the full load of home and caregiving responsibilities on his own. I've been so fortunate to have a partner in life who not only is a wonderful father to our children but also has been supportive of my career. My progress thus far wouldn't have been remotely possible without his steadfast support. I've often wondered what home life looks like for women who don't have Marcos in their lives.

In addition to pipeline problems, sexual harassment, and the boys' club and its nepotism, there's also crazy internecine scientific vendettas and bullying in the academy. Several of my mentors—men and women— have endured career-altering conflicts with an academic nemesis. Driven by intense competition for scarce funding resources and disagreements over intricate nuances in their respective fields, these toxic relationships can undermine scientists in their pursuit of funding and even hinder the publication of scientific results. In some cases, this can hurt students and trainees, stymieing their progress for years to come.

I experienced this form of academic bullying a few years after I finished my PhD, during the opening reception of an international conference held in a beautiful old European university gallery hall. Renaissance paintings decorated the walls to the great room, and marble busts of famous historic scientists and scholars were staggered down the great hall, a football field in length. Standing alone at a high-top cocktail table, enjoying some of the dainty hors d'oeuvres passed around by the waitstaff, I was approached by Professor Snake, a tall man eight to ten years my senior, with graying hair and beady brown eyes.

After exchanging the briefest of pleasantries, Professor Snake began to forcefully tell me how he thought that my latest review paper written with Andrea and other collaborators was utter shit. His face turned red, and his body was visibly shaking. A stream of sweat ran down the side of his face. He was enraged that we hadn't cited one of his papers—*which didn't actually fit our review criteria*—in the article.

I stood there, mute with shock, frightened by his demeanor.

I'd never even heard of someone going up to another scientist out of the blue to complain about a reference citation in a paper before. It was a disturbing encounter and not the last for me or for others. Professor Snake has gained a reputation for bullying and harassing women scientists, whether in vocal confrontations or in blocking their papers in his role as a reviewer and editor, or writing toxic commentaries on their work. He's also done this to my papers and gloated about it to my colleagues.

Most disappointing has been the response of the scientific community to his behavior. When I confided in a senior professor I looked up to, his response was a litany of excuses for Professor Snake's behavior. *He has a difficult marriage. He's having problems with his job. Can't you be more empathetic? You must be overreacting.* Was that supposed to make me feel better? What about my students whose papers are getting rejected by him just based on their association with me? Were the other victims of his bullying and I expected to continue to take this abuse?

When I reported his behavior to the executive committee of the scientific society that hosted the conferences where these caustic in-person encounters took place, they listened to me . . . and then did nothing. I requested that they implement a code of conduct for future conferences to discourage harassment. They refused.

Professor Snake has yet to face the consequences of his actions. Sometimes there is no other solution than to walk away from both the organizations and individuals that shield the behavior of the Professor Snakes of the world.

If there is a silver lining here, it is that this experience led me to advocate for inclusion of codes of conduct in other proactive and supportive academic societies that I'm affiliated with, such as the Society for Economic Botany. It is heartening that there are now clear guidelines on what is considered unacceptable behavior at society events, a reporting structure to use, and clear consequences for the offender that can be used in these situations in the future.

All this bullying brings me back to the days of growing up different. I had one or two surgeries a year throughout most of my elementary-school years. Those pesky growth spurts resulted in bone spurs that had to be sawed off and my femur length corrected. The weakness in my leg from the six-month-long bone-lengthening procedure required that I use some extra support. After I spent months on crutches, the doctors wanted me to put weight on my right leg to build up strength, so I was assigned an ugly metal walking cane. I was nine years old and in fourth grade when the real teasing and trauma began. I never cried in public, always at home in bed by myself. These pity parties were a sad state of affairs. The name-calling is what got to me the most. Some would call me *crip*, but one boy in particular delighted in tormenting me.

Every day, he would chant, "Granny, Granny . . . here comes Granny with her cane!"

It was always worse on the playground, out of earshot of the teachers or classroom aides. Then, one day, something just snapped inside of me.

During one of his singsong taunts, I lifted up my cane and whacked him on the head with it, shouting, "I'll show you who's a Granny!" He fell to the ground with a dramatic thud. In retrospect, I must have hit him quite hard in my pent-up rage.

One of the teacher's aides ran over and asked what happened. I looked her straight in the face and said, "Billy fell off the swing."

He never called me Granny again and didn't tell the teacher. I was never punished for what I did. A part of me thinks that perhaps she saw what happened and let it pass. Who knows. The lesson here, though, is not that I found violence to be the answer to my woes, but I do think that this moment was a turning point for me. It was the moment I had my first taste of empowerment in place of weakness, and it felt good.

In 2016, just a year before my encounter with Professor Toad, the Phytochem lab had been filled with a five-person film crew. We'd scrubbed the lab spotless before they arrived, and several students from the lab team were waiting outside in the hallway, anxious and excited for the opportunity for their walk-on roles.

"I want you to walk down the hall, and we'll follow you with the camera," Peter, the field producer, explained.

"Are you sure I need to walk for this?" I asked nervously. "Aren't the still shots of me standing and speaking in the lab enough?"

"No, no. We need to show you coming in to work. Just a short clip to orient the viewer to the building and the lab, and then I need some shots of you bringing the plant samples from that bench to the machine over here," he responded.

"I need to adjust your mic. Do you mind?" One of the audio technicians indicated for me to lift my lab coat and shirt as he taped more of the wiring in place.

The lab was filled with high-end filming gear, high-resolution cameras, mic booms, and lighting stands. James and Kate waited in the back corner of the lab, ready to make their own appearances on the show when the time came. The film crew was working on a new TV series that would air on the National Geographic channel called *Origins: The Journey of Humankind*.

The crew had flown out to Atlanta from Los Angeles to film me for the episode on the history of medicine. It was a huge honor to be featured, but I was battling some serious insecurities. Just months prior, I'd been photographed with my leg on display for Ferris's story in *The New York Times Magazine* detailing my search for new antibiotics from medicinal plants. To simply have my prosthetic on display hadn't bothered me, but showing the world my awkward lopsided walking gait was a whole other matter.

I experienced a painful flashback to the eighth grade when I'd had the *marvelous* idea of having Daddy tape me working on my science fair experiments in the hospital microbiology lab with our big VHS camcorder. I was determined to keep up with the rich-kid school districts and have a video of my work on display with my science fair board at the state competition that year.

In preparation for the filming, I carefully prepared my hair in the style that was popular with thirteen-year-old girls at that time. It involved long bangs, a hot curling iron, a brush, and enough Aqua Net hair spray to light a large fire had I gone near a flame. The result was a double-curled, poufed-out, four-inch-long knot of hair frozen in place on top of my forehead. Once in the lab, I donned my knee-length white lab coat and latex gloves, and then began explaining the process of my experiments to the camera. This involved me walking back and forth across the lab from incubator to biosafety cabinet as I went over each step.

My limp from my prosthetic was especially evident and, combined with my worsening scoliosis at that age, had the effect of making me appear hunched and creepy. Add to that my pale skin, bloodred lips, and big bangs exploding from my forehead, bedecked in a lab coat plus a mouth full of orthodontic braces, dragging my fake foot along the floor as I beckoned the camera with a long wave of my hand to "follow me into the lab," and I bore a striking resemblance to Dr. Frankenstein's mad assistant—Fritz, aka Igor—from the classic 1931 horror movie.

To make matters even worse, Daddy had grabbed one of his wrestling tapes (he was a big fan) the day that he recorded me in the lab, and had taped over the first part of the cassette with my lab scene, but the rest of the videotape remained high-octane wrestling-ring antics. I didn't realize this, of course, until I was speaking with a judge at the state science fair competition with my video playing in the background. When the judge's jaw dropped open as he looked at the video behind me, I turned and, to my mortification, saw (and heard) a World Wrestling Federation character *pile drive* his opponent in the ring. He then proceeded to lift and shake his arms with an aggressive roar before strutting around the ring, bare chested, wearing only his tight neon-yellow micro-shorts and bright blue thigh-high boots.

And now they wanted to film me walking into the lab—with my persistent limp, though with more tasteful hair and makeup—to be aired on the National Geographic channel. My gut clenched up and I battled the urge to vomit.

I talked it through for a bit with Peter, and in the end, I did it. Despite my desire to portray myself as a normal scientist, I was anything but. Maybe it was important not only to show the audience that I could do the work I do, but also to show the truth of my imperfect, lopsided body, which moved without a modicum of grace.

The next day, I jumped on a plane with the whole crew to fly to south Florida. Traveling with a team and getting out in the woods helped me loosen up, and we had some fun spots to hit. We explored swamps and oak hammocks and even took out an airboat. I learned important lessons along the way and got over some of my self-consciousness. I also took the opportunity to collect and fill my suitcases with a plant we were currently working up in the lab: the Brazilian peppertree. This species especially loves the large drainage ditches that border pastureland. It's included in the list of noxious weeds in Florida, meaning that its cultivation is prohibited. Luckily for our research, there are abundant wild populations to sample from.

As my lab team grew, so, too, did the number of undergraduate research projects, and this presented me with a unique opportunity to assign my currently unfunded projects to students to generate more preliminary data. One such student was named Amelia. She was very driven, balancing her time between her studies, work as an emergency medical technician on campus, and research in the lab. She'd spent the past semester training with Kate in various lab techniques ranging from work on plant chemistry to antibacterial assays with fluorescent quorum-sensing reporter strains of staph bacteria that Alex had provided.

The day I returned from Florida, I dragged my suitcases loaded with peppertree into the Phytochem lab. Kate, James, and Amelia quickly got to work helping me sort the different parts into three buckets: leaves, stems, and fruits. The fruits were beautiful, and the clusters of vibrant red berries quickly filled our white work buckets as we clipped and sorted the samples.

While it may be a hated species in Florida due to its invasive tendencies and resilience to insects and pests, the Brazilian peppertree has a fascinating history of medical uses. As a part of the background research for her project, Amelia dug into scans of texts from the 1800s in her search for details on its medicinal applications, which included uses of the fruits to treat wounds and ulcers, while the bark was used to treat vaginal, urinary, and skin infections as well as burns, and the leaves were used for eye infections, wounds, ulcers, and rheumatic pains. Scouring the pages of the 1648 *Historia Naturalis Brasiliae*, written in Latin by the Dutch naturalist Willem Piso, she found some of the oldest known accounts of its use as a medicine. While the leaves and bark had been well studied by a number of other scientists, the fruits had not. And so I assigned her the task of further examining the ability of the fruit extract to block staph quorum sensing, and to refine the compounds responsible for the activity through teasing apart the complex medley of chemicals in the lab. Now that we had new tools in place in the lab from our collaborative work on chestnut with Alex, we were able to look for other groups of compounds in very different plants in our hunt for novel virulence inhibitors.

After those early years of ramp-up, my lab was finally running with a critical mass of students, trainees, visiting scholars, and research staff that merged their complementary expertise across the fields of botany, microbiology, pharmacology, and chemistry. We were making leaps forward in our scientific progress, putting out research papers and grant applications like a well-oiled machine. This progress came at a cost, though—mainly one of time and my ability to manage it. I struggled with juggling my various roles as lab team captain, professor, mentor, mom, wife, author, editor, reviewer, and grant writer. I also continued to run up against frustrating barriers to the advancement of my career.

ACADEMIC SERVICE, RESEARCH PAPERS, and invited talks are important to career advancement, but the true academic currency that determines whether or not you are promoted, earn tenure, and experience upward mobility is this: your ability to bring in extramural research grants that carry overhead funds to the university. The dollars brought in by research enterprises are a big source of funding for academic institutions, and most of the pressure around the success or failure of academic scientists is tied to money.

Now, imagine competing for these resources at a disadvantage. Your every activity—whether a research paper or a grant submission—unduly scrutinized for your "ability" to do the science you are proposing; your expertise, findings, and qualifications always in question. Will you be judged fairly? Will your project be dismissed as "overly ambitious" for your skills? Can you dare to hope that your ideas will be considered innovative, potentially paradigm shifting? Or is it more likely that they will be written off as an outsider's amateur efforts, and ultimately squashed for daring to creep outside the dogmatic box of what Western science should be?

Sexism and implicit bias affect scientific progress in unimaginable ways. Research has shown that women and minorities face waves of implicit—and sometimes explicit—bias at every level of the science enterprise today, from reviews on grants and papers to hiring and pro-motions. They are often underpaid in comparison with their peers and receive much smaller development packages—start-up funds provided by universities to kick-start a lab—than their white male counterparts. This can have long-standing impact on the success or failure in the com-petition for extramural funds.

Imagine this scenario: if you were to pair two newly hired faculty and compare them, one receiving 40 to 50 percent less in start-up develop-ment funds, which of the two is going to be able to generate more of the necessary data to successfully compete for research grants? Clearly, the one with the funds to hire more scientists on their team or purchase nec-essary supplies. By shortchanging women and minorities in start-up packages—the dollar amounts of which are typically held secret—institutions set them up for failure on day one. Once someone is on this path, it is incredibly hard to renegotiate and secure equal funds. Often the only recourse is to secure a job elsewhere, requiring the faculty mem-ber to uproot their family and placing tremendous stress on individuals already pushed to the limit.

Tenure-track PIs not only teach, they also lead research teams (some large, some small), train graduate students, and manage multimillion-dollar research budgets. In essence, each PI is the head of their lab do-main, responsible for leading other PhDs and PhD trainees in successful research endeavors. This is where few women and minorities dwell, and even fewer make it past the gates of tenure. It can be an isolating and daunting place to be. I know this feeling well. Until just last year (2020), I was the only woman PI in my office suite of twelve faculty, the only woman with a lab of her own on our entire research floor.

While things have certainly improved since the time of Dr. Rosalind Franklin in the 1950s, nearly eighty years later, women still do not stand on equal ground. The goal mark constantly shifts in the game of science, and to succeed as someone who is "other," I've learned that I must work harder, push harder, than sometimes feels humanly possible. That is what it has taken to keep up and hold my spot in this race.

The worst and most enduring experience, for me, at least, is the consistent underestimation of what I'm capable of, even disbelief at what I've accomplished. I often feel like I have to walk the line of wanting to say exactly what I've done but worrying that I will sound pompous or risk being labeled a braggart. At the same time, I feel a persistent need to gird myself with the armor of my accomplishments, lest I be pushed aside. Humility is a luxury that only white men in my position can afford.

I began my journey in science as a child. It was my sport and my joy. It continues to be those things to me as an adult, and I cherish the moments when I get to share my love for the scientific process and awe for exploration and discovery with my students and trainees. Only now I understand that science is also a blood sport; in the major leagues, only the toughest survive. This isn't just an issue about men versus women. It's about power—and how those who hold it wield it.

BARRIERS AREN'T ALWAYS FORMED by others. At times, I've suffered from obstacles of my own making. I tend to be generous with my time and review papers and grant proposals of my peers, but I hate asking people for help. It makes me feel uncomfortable to place my burden on them, whether they are colleagues or trainees. Perhaps this is the result of a life struggling for physical independence that has shaped me to feel that asking for help or feedback is a form of weakness. A flaw that others could exploit.

Just a few years into my career, I began to sense that my desire to keep others happy was hurting me, both physically and mentally. I wasn't eating healthily or getting enough sleep or exercise. I wasn't able to spend the amount of time I wanted to with my family. My workdays got longer and longer, and after a full day in the office and lab, then dinner and bedtime for the kids, I was back at it until the late evening. My service load across university committees, academic societies, editorial boards, and grant review panels was unbearable, crippling to my research productivity. I began to understand that each time I said yes to one more request, I was actually saying no to my health, no to my family, no to my own research success. I needed help! How could I prioritize? How could I say no, even to small requests, without causing insult or disappointment? Each small request was one more piece of straw closer to breaking my camel's back. How should I respond when others pressured me to change my mind even after my initial no? *Would saying no more often hurt my career or jeopardize my path to tenure and promotion?*

I turned to women in science group meetings for advice in both the college and the medical school. What I found was that these struggles were shared. While there was some comfort in learning that I wasn't alone, it didn't solve the problem. We itemized complaints and brainstormed solutions, but nothing ever changed for us. We didn't have the power to *make* changes. We battled issues with access to childcare near campus; salary disparities and the financial burden it placed on our households; overloaded service commitments, scheduling, administrative duties; and the emotional weight of the many students whom we spent hours counseling through their struggles. We noted that many of our male colleagues avoided the service roles we took on and were instead cloistered in their offices productively writing grants and papers. Like iron balls shackled to our legs, these other commitments were dragging us down, and no matter how hard we tried to climb, we kept getting pulled back down again.

I was demoralized and decided I needed a new vantage point: I wanted to understand how the other half lived and worked. Maybe I should see how my male colleagues dealt with their challenges. We were, after all, in the same jobs. Even while there were disparities, we all had only twenty-four hours in a day. I decided to do some reconnaissance.

HEY, DUDE, CAN YOU pass me a beer?" I asked Steve.

We sat crowded around a dining room table in Joe's house, in a suburb of Atlanta. We'd begun meeting one night a month, moving our gathering from house to house as we shared in the hosting role, providing simple fare of tacos or pizza and beer . . . sometimes capped off with a good glass of Scotch.

In the room sat some of the top microbiology professors in town, a ragtag bunch of transplanted Brits and US-born scientists who taught and led research groups at Emory and Georgia Tech. The group was made up of mostly men and one other woman besides me who all gathered to pitch grant ideas and strategize on how to raise more research funding and organize larger-scale gatherings and scientific symposia. We shared common interests in the topics of infectious disease, evolutionary modeling of antibiotic resistance, and drug-discovery approaches to tackle killers like cystic fibrosis and sepsis.

In Victorian times, the educated and wealthy elite held salons in the home, where a select few guests were invited to attend and discuss the latest topics in science, culture, and politics. I've always been entranced by the concept—I love to host a good party, and especially one with interesting conversationalists. We weren't rich, and certainly not elite. While politics sneaked into our conversations sometimes, most of the time it was pure science and strategy. Many of the regulars ran hugely successful labs, and I relished these evenings as learning opportunities for ways to up my own game in the field and learn from their success strategies.

That evening, we discussed a topic of great frustration to all of us: the grant writing process. Like just about every other female PI I knew through my women-in-science networks, I wrote all of my own grants myself, and sometimes would ask my team—mainly the more senior grad students and postdocs—to provide feedback, even though I felt guilty doing so. I felt like I was taking advantage of them when I did this, because even though the funds would ultimately be used to pay for their salaries and research supplies, if I were successful, the credit was awarded only to me. I also never asked my colleagues for help. If they weren't co-investigators benefiting from the project, it felt wrong of me to do so.

We went around the table, each person providing updates on their latest applications and what had gone wrong or right in the process. When it was my turn, I explained my writing process. Their shock was evident. Mouths hung open, glasses went untouched, uneaten pizza cheese congealed on plates.

"What?" I asked.

"You're writing your own grants—all by yourself?" Marvin, a microbiologist, finally asked.

"Yes, of course. You don't?"

There was silence. And then everyone said, no, they didn't. I was shocked.

"Cassy," Marvin said, "you're actually doing your team a *disservice* by not having them more involved in writing the grants. How are they supposed to succeed in science if they don't know how to write a grant proposal? I divide my team into small groups and they work on the proposal on their own, with my regular feedback. I think you should try it."

Marvin had brought in millions of grant dollars to support his research program and trained many students and postdocs, and those students and postdocs actually had a huge rate of success in landing jobs and launching their own careers after finishing in his lab.

The others in the room explained that they, too, used similar strategies. No one wrote their own grants single-handedly, and without peer feedback, like I did. They explained that when it was time for their team members who had led initiatives on grant proposals to apply for a job, they always gave them huge credit for their work in strongly supportive letters of reference, emphasizing how they took leadership in the grant application process, ultimately helping them land the jobs they deserved. They didn't feel any guilt over the process. In their opinion, this method was a critical part of scientific training.

Was this a mistake that all women in science made, or just me?

I needed this direct feedback—and this shift in my perspective. I'd felt that I needed to do all my own work on these applications to prove myself and not unduly put work on others. But my colleagues helped me see that this was not only too taxing for me but also, as important, a missed opportunity for those on my team. I should empower them to help me: they would need to know how to write grants down the line. This was a win-win. My colleagues also encouraged me to reach out to them for peer feedback on future applications. They explained that this is how we could lift one another up, and elevate our collective impact on the field together. I left that meeting determined to change.

Around this time, in my late thirties, I made another change. I moved away from the flesh-colored foam carvings that gave shape to my calf muscle in attempts to replicate the look of a real leg. Those fake skin covers made me feel inadequate and phony, a mirror reflecting my inner turmoil over my place in science. With a second glance, anyone could see that it wasn't a real leg. I was tired of pretending.

I went to my garage. There, I have a box of old legs. Some are still in good shape. Others have not fared so well, the foam cover or the plastic foot nibbled on by rodents during the years they were left out in the barn behind my parents' home. I lined up all twenty, from shortest to tallest. This battalion of legs would probably look morbid to most people, but for me it told a story. Each leg was its own chapter, its own era. Little soldiers standing at attention, each one represented a time of battle in my life when taking just one step forward required that I first overcome significant hurdles.

My legs have changed over the years, transitioning from heavy, wooden pieces with lots of straps to lighter metal frames paired with tight braces. Now they even have ankles that allow a rolling motion. I still limp wearing them, but my gait is closer to normal. When I was a kid, having each leg built was a multistep process, requiring trips back and forth to the prosthetics office. That's where I met Charlie. First, Charlie created the cast mold of my stump, which would later be used to build the test socket. He would slide a thin cotton sock over my stump and pull it tight. Using his knobby old hands, he pressed firmly into different parts of the stump, drawing a line where the top of my tibia sat, a circle around my patella (kneecap), and so on. Then he brought in an old bucket filled with warm water. He opened plastic bags filled with tightly bound rolls of cotton strips—about the size of an Ace bandage—that were coated in dry plaster. One by one, he dropped each bundle into the water and then reached inside to massage and squeeze it with his hands beneath the water's surface, wrapping the hot plaster strips around my stump.

Charlie sat on a stool in front of me, his crinkly head of gray hair tickling my chin as he bent over at his work. Never content to sit in silence, I peppered him with questions about the process, and he would patiently respond.

"Why does the plaster get hot?" I asked.

"There's a chemical reaction it has to complete to turn hard," he replied.

Once all of the plaster had been applied, he wrapped his hands around the stump, using his thumbs to put pressure on the points where my stump would eventually bear weight inside the prosthesis. He held that pressure and the shape of the cast until it hardened. Next, with the skilled flick of his fingers, the clips attached to the belt at my waist would be released, and the cast was carefully removed. Inside the cast, the purple markings from the lines drawn on my stump bled into the plaster, and could be seen to serve as a guide as he created the final socket mold.

Charlie was an artist. And, though it sounds strange to say, he was also a healer. He handled my stump like it was part of his art: he didn't look at it as an abnormality, a shortcoming. To him, I was no freak to be teased. I liked how he handled my stump—it was just a *thing*. Maybe it was easy for him to do that work, as it was his job after all. But I drew courage from Charlie's attitude when I finally decided that I wanted a change.

I drove over to see Will Holbrook at Fourroux Prosthetics, my leading leg man throughout adulthood, and told him, "Just strip it bare. A metal stick is all I need!"

I wanted people to see the reality, to show them I was done pretending to be something I wasn't. If they were going to gawk and stare, let them stare at *that*!

Will took in my dramatic declaration with a smile and an idea.

"You know, Cassy, there are some new 3D-printed covers being made now by a couple of companies. There's one in Spain, one in Canada. Check out their websites and let me know what you think."

Each site had colorful prosthetic covers in various designs. There were floral motifs, others featuring birds and turtles, or even robotic-style linear patterns that reminded me of *Blade Runner* or *Brazil*. Scrolling through the websites was more like shopping for art than for a medical device. I was hooked.

Working with Will, we made the measurements—the length of my metal pylon from the artificial ankle to my kneecap and the diameter of my stump socket. I would be the first of his clients to try this out, and we were both giddy in our excitement when the box arrived in the mail. While Donato lobbied hard and eloquently for the covers that had superhero-style decor—Spider-Man, Star Wars Stormtrooper, and Iron Man—my first choice was a gunmetal-gray pattern named the Victorian by the Canadian company Alleles. It had a steampunk vibe that I loved. Not long after being outfitted with this new prosthetic, I noticed a change in the gaze of strangers. No longer were their stares full of pity; instead, they evinced a measure of awe. What had once been a source of shame and uncomfortable questions became a fun talking point with friends and curious strangers alike.

The cover was a piece of art, a reflection of my mood of the day, and a sign that I was finally starting to open up and accept myself in all of my differences from the norm.

CHAPTER 11

The One-Legged Hunter

I was taught to strive not because there were any guarantees
of success but because the act of striving is in itself the only
way to keep faith with life.

MADELEINE ALBRIGHT, *MADAM SECRETARY*, 2003

In midsummer 2019, I was leading an expedition across the Sharri Mountains of Albania and Kosovo, riding shotgun in an SUV as it crawled over yet another large stone on the rough trail en route up to the national park. While my old and ongoing collaborator Andrea wasn't with me this time, I was here continuing the work that he and I had begun in 2011 as part of our interest in tracing the Albanian origins of some of the ethnobotanical traditions we'd documented among the Arbëreshë diaspora in Italy.

On this trip, I was working with our long-standing collaborators at the University of Prishtina in Kosovo's capital city. In addition to the local team made up of Professor Avni Hajdari, chair of the Department of Biology; Xhavit, a scientist and park ranger who knew this area better than anyone; and Bledar, a graduate student in Avni's lab, I had also invited Susanne Masters, an orchid expert from Leiden University in the Netherlands. Rounding out the group were Spencer and Jon, *National*

Geographic photographers. Spencer was behind the wheel of the SUV, having said he had experience with stick shifts and rough terrain. Although I don't think he—or any of us, really—realized just how rough the roads would get. As we climbed higher up the rough road, our dark green seven-seat Land Rover SUV strained with effort. The smell of burning rubber permeated the air; we were grinding the clutch in our ascent.

Eight years after initially visiting for field research, I was back in Kosovo for a project funded by the US State Department under its scientific capacity building initiative. My primary collaborator was Avni, and our continued work in the country was due in large part to his high level of research productivity and perseverance in securing funds to expand the reach of his department and university. We had two priorities on this journey. The first, as always, was to collect species of interest for further study, mainly wild yarrow (*Achillea* spp.) and medicinal succulents. But the bigger and even grander design was for me to help establish a microbiology laboratory to enable local scientists to test extracts of medicinal plants found in their country for antimicrobial activity. It was a dream come true. This would be a hugely important step in expanding their ability to tap into their own abundant natural resources. If they had their own lab running their own tests, we could speed up the overall ethnobotanical approach to drug discovery. This was the first lab I'd been part of building outside Emory, and it proved that there was value in my way of doing things, not only for my own lab but, more important, for the people who lived in this region, whom I had learned from over the years and who would soon be able to make their own breakthroughs in finding promising antibiotics in plants.

A newly formed nation, not yet recognized as a sovereign state by all countries, Kosovo rose from the ashes of the Balkan conflict of the 1990s. The ethnic Albanians and the Kosovo Liberation Army were at war with the ethnic Serbs and the Yugoslavian government (comprising the republic of Serbia and Montenegro). In response to the uprising, the Yugoslav and Serbian forces, led by Serb president Slobodan Milošević, engaged in a program of ethnic cleansing. The United Nations condemned it and NATO intervened, with the United States playing a prominent role. During the horrific conflict, more than 1.2 million ethnic Albanians were displaced and at least 12,000 people died, including unarmed women and children. The Serb military, paramilitary, and police committed acts of sexual violence and rape on an estimated 20,000 Kosovo Albanian women. The war ended with the Kumanovo treaty in June 1999. Years later, the International Criminal Tribunal for the Former Yugoslavia charged Milošević and several commanders in the Yugoslav army for war crimes and crimes against humanity.

Twenty years after the conflict ended, NATO still leads peacekeeping-support operations known as KFOR (Kosovo Force), which consists of 3,500 troops provided by twenty-seven countries. On my first visit to the capital in 2014, I was surprised to see a statue of former president Bill Clinton in the city, and roads named after George W. Bush. The Kosovar people credit their freedom and rescue to the United States and NATO, and they're proud to show it. This country has been through hell, and yet everyone I met was extremely generous. There is no country on earth where I have felt more appreciated by complete strangers than this one.

We'd learned of the medicinal uses of a fleshy succulent plant—a *Sempervivum* species from the Crassulaceae family, commonly known as a houseleek, or hens and chicks—that locals used for ear and sinus infections.

Anjur, a local man knowledgeable of these things, had pulled off a single fleshy leaf and given it a good squeeze, and a clear, slimy liquid dripped out. "You put this in your ear when it hurts, or your nose for sinus infections," he'd said. Anjur grew it in a small pot in his courtyard garden but explained further: "It grows up high, on rocky peaks in the mountains; that's why I brought it home to grow it in a pot."

This was a species we'd never looked at in the lab, and based on its dual uses in treating ear and sinus infections, I had a hunch that it could have some activity against gram-positive bacteria like *Streptococcus pneumoniae* or *Staphylococcus aureus*. Our ace botanist and human GPS, Xhavit, assured us he knew where to find this plant. "I know where," he said in Serbian, his eyes lighting up as he wiped his wizened brow and brushed his white hair back under a cap. "I know where."

My favorite three words in the English language are when Marco says to me: "I love you." My second-favorite three words are when anyone, anywhere, replies, "I know where," when asked where a plant I'm looking for is. I always get giddy.

While the road was rough, the view was spectacular: wild and rugged, countless hilly meadows covered in multicolored wildflowers, as if a rainbow had sprinkled a blanket of color across the ground, majestic mountains in the distance. We stopped a few times to trek through the muck of soggy meadows to document terrestrial orchids spotted along the way, collecting ones available in large populations for the Prishtina herbarium records, and photographed and collected a single flower from the rarer species.

Storm clouds appeared in the distance, and lightning struck in a far valley. I didn't even hear the thunder, but I began to worry. The wind could blow the gathering darkness in our direction. Our road was bad even when dry. A downpour would leave us stuck here for the night or even longer. We'd driven past the dwellings of rural farmers miles back down the trail where we could seek shelter from the coming storm, but it would be slippery and dangerous to make the trek in the rain by foot or car. I asked Xhavit how much farther it was. "Not much longer," he insisted. I trusted him. As expedition leader, I made the call to continue.

It took us another fifteen minutes of crawling the car along the rough path until the path ran out. Not far ahead, we could see a huge cliff face rising hundreds of meters upward from a valley below. Most of the mountain facade was bare and gravelly. We turned the car and left it pointing down the narrow strip of dirt road so that if the brewing rainstorm landed, we wouldn't have to try to turn our car on a slope by a precipice in slippery mud. I watched the far valley, and the clouds didn't seem to be moving in our direction.

By far the eldest member of our team, Xhavit, with the help of a wooden walking stick, took off on foot, racing up a grassy hillside to make his way to the rugged mountain. Used to this routine, Bledar hurried after him, carrying his camera and collection bags, while the rest of us explored the grassy wet hillside. After slowly hiking to the top of the hill with my gear and walking stick, I glimpsed Xhavit and Bledar far, far ahead; they looked like ants scurrying along a ridgetop. Tired after the climb, I plopped down to rest in the grass and took a moment to admire the vast wild landscape that surrounded me. It was all so verdant and alive, earthy with the sweet scent of freshly crushed herbs where I was sitting. It was peaceful and beautiful and all of those things that make you want to sear an image—a tremendous moment—into your mind for eternity.

Susanne and I got to work hunting for species of interest to add to the herbaria collection or for further laboratory study. We found not just one species of lady's mantle (*Alchemilla* spp., Rosaceae), but three! Then there were leafless broomrape (*Orobanche* sp., Orobanchaceae) that have no chlorophyll; instead they parasitize other species for energy and nutrients. I lay down in the grass, using my camera to snap shots of this floral bounty, before snipping off specimens.

A great shout boomed in the distance. I saw Xhavit charging up the hill with a large bag in hand, Bledar trailing behind with another. Xhavit had the most striking smile when he waved up to me, beaming, motion-ing for us to come, look at his find. He reached into the cloth bag and with pride lifted up a gorgeous cluster of a unique houseleek to the team. Mission accomplished! He had found it! Beside him, Bledar showed off his finds as well, which included several species of yarrow (*Achillea* spp., Asteraceae), as well as a hard-to-find endemic one. Bledar's doctoral dis-sertation was focused on yarrow, so this adventure had been very fruitful to his research. It was this kind of international teamwork I'd grown to really love; it didn't matter where we came from, we were all studying similar things and hoping to achieve the same ends.

And at that moment I felt that familiar swirl of dog-tired fatigue buoyed by sheer joy and excitement. It was like my body wanted to col-lapse but it was filled with helium. We found our target plants, and the photographers had captured amazing pictures of the flora and of the region itself. Maybe it was the adrenaline, or maybe the exhaustion, but the drive back down the mountain path didn't seem nearly as scary as the way up had been.

After a hearty dinner of qebapa (small grilled ground meat—a mix of lamb and beef) and speca me mäze (Albanian peppers in cream sauce) served with fresh pita bread, it was back to work at our hotel in Prizren until midnight. The hotel manager had been kind enough to allow us to spread out our gear and samples in the hotel restaurant after the last patrons left. We needed to catalog and process all our new specimens before they could fall prey to natural decay. We divided up the tasks, checking the plants with our books, typing up the data, and ensuring that each specimen was properly arranged in the press. As we didn't have permits to bring the CITES-listed (short for Convention on Inter-national Trade in Endangered Species of Wild Fauna and Flora) orchids back to the Emory Herbarium for digitization, Spencer helped us docu-ment them with super-high-resolution photos for our records. All of the physical specimens would remain in the University of Prishtina herbarium. After this day, as the day actually became the next, I took a hot shower, took off my leg, and fell into a bed that had never felt so good.

LATER THAT WEEK, in Avni's chemistry lab at the University of Prishtina, we sorted our dry plants, splitting up the samples; some would be shipped back to Emory, the rest remaining for his students to extract for essential oils and antimicrobial analyses. Over the past few months, we'd been in communication to strategize over which pieces of equipment were essential for the new microbiology lab and where we might find the best prices. We ordered an incubator, a plate reader to measure the optical density of bacterial cultures, sets of pipettes, and a centrifuge. The costs added up, but the funds from our latest capacity-building grant from the US government were just enough to cover the expenses of setup and leave room for the purchase of the supplies (petri dishes, growth media, test tubes, and antibiotics to use as controls).

Avni is a few years my junior, and always has a smile at the ready. He loves to tell jokes—the types of jokes that are so corny that you can't help laughing with him. He and Bledar had already spent months training with my research group back at Emory two years prior, and Avni's lab was set up to run their own analytical chemistry tests on the essential oil extracts derived from local Balkan flora; they were eager to expand their research tool kit to include these important antimicrobial tests. In advance of my arrival, Bledar had unpacked the supplies and set up the equipment in the small lab, a room newly dedicated to these efforts at the university. Two additional biology students joined us in the lab to run a demonstration on how to test their plant extracts for antibiotic activity. Bledar took the lead, and I sat back to observe and answer any questions they had along the way. The lessons from his time training in my Micro lab had stuck with him; he moved through the experiment with an ease born of practice and confidence, describing each step to the students in Albanian as he worked. Is there any better moment for a mentor than when we see that our mentees don't need us anymore? Obsolescence is the nectar of mentorship.

Sitting in on a corner stool, dressed in a borrowed lab coat over my T-shirt and jeans as I watched them set up the experiments, I took a moment to reflect on how we'd gotten here. Avni and I shared our first joint research publication eight years ago in 2011 when we were both still postdocs without labs of our own. So much had changed in the years since, and we'd both stayed busy building our respective groups, training students in the art of research, and teaching classrooms full of undergraduates—5,400 miles apart.

When I'd had my first big break with the large research award while a postdoc at Emory seven years ago, I'd dreamt of a day when my lab could do exactly this. My vision had been to build and lead a research group that would not only make scientific breakthroughs in our lab, but also work in concert with a network of collaborators around the world, expanding research access for all parties involved. With more than thirty-three thousand medicinal plants spread across the globe, there was much to investigate, and when it came to battling antibiotic resistance, a united front and an army of scientists were what we needed. It was an incredible thing to see that vision begin to take shape. One by one, through ethical engagement with local communities and long-lasting collaborations with scientists and their students in biodiversity-rich countries, I knew we could make a difference.

AFTER WRAPPING UP OUR work in the lab and herbarium, I wanted to visit some of my old Gorani friends living in Albania at the highest elevations of the Sharri Mountains. Avni suggested we also visit a village on the Kosovar side of the border, but I vetoed that idea—and not without reason.

A few years ago, we'd visited that community with Andrea and a team of students. At the time, something had struck me as being off in that particular village. Although part of an ethnic minority group in the region with unique cultural practices, it was an outlier from other communities of the same ethnic group. Men and women were not only kept separate, but were even required to use different roads throughout the village—neither party was allowed to travel on the other's. What worried me most was that even the elder women, who'd indicated that they wished to speak with me and my female interpreters, were fearful of being caught speaking to us—even on a topic as innocuous as food plants.

Later, we'd heard rumors that some of the older teenage boys and young unmarried men had been radicalized and left home to fight in the Middle East. In impoverished regions where young men have few job prospects, the promises of riches with ISIS were too great to turn down for some. I had no desire to return to that particular village. It was too dangerous both for us and for the few locals who would even agree to speak with us.

"Let's just go to Borje in Albania. You'll love it there—the views are incredible in the mountains," I told Avni. I'd remained in touch with members of the Borje (pronounced *Bore-ee-yuh*) community using social media and was eager to see some of them after many years away. Although they had limited access to common market goods, there was an internet café in town and many of the young men used Facebook on their cell phones.

Andrea and I first traveled there together back in 2012, along with our extraordinary interpreter, Mezahir, who spoke not only fluent English and Albanian but also the Gorani language, a Torlakian transitional dialect in the Bulgarian-Macedonian and Serbo-Croatian languages. Notably, Mezahir's day job was radio DJ, and he was a natural when it came to interviewing people. While I normally undertook my own interviews in Italian or Spanish, work in the Balkans presented unique challenges. It wasn't unusual for our research team to interact with people of distinct language groups in different communities in the same week, and while I'd made some early progress in mastering some of the basics of Albanian, Serbian and Gorani remained elusive.

We debated whether to try the mountain border crossing—the closest option to getting there from Prizren—but I'd learned my lesson from Andrea's past attempt at the same, which nearly landed him and two colleagues in a rural Albanian jail. "No, we'll take the long route through Kukës," I told him. It would add an extra hour to the drive, but we'd use the official border checkpoints required of foreigners.

I sent out some messages to my Gorani contacts on Facebook and helped the team load gear, along with some sandwiches made with fresh bread and filled with sheep's cheese, into the SUV. The border crossing was simple and the highway led us just outside of Kukës, the first city ever to be nominated for the Nobel Peace Prize for its hospitality in embracing the 450,000 Kosovar Albanian refugees that fled there during the war.

Turning off the highway, we made our way up gravel and dirt paths rutted out by the rains and movement of livestock from village to village. A stream of crystal-clear mountain water flowed in the ravines below the road, the water crashing against large boulders and small gray river rocks as it meandered downward to Kukës. It was an unusually sunny day; I'd never seen such a vivid blue sky there before. Cumulus clouds floated above, as if daring us to take pictures of them hovering around Mount Gjallica—a limestone mountain covered in lush green coniferous forests, the highest summit in the region.

The other mountainsides were a mix of forest with expertly positioned terrace gardens carved into the steep inclines, where locals grew a mix of seasonal vegetables and potatoes. Though the potato has its origins in similar montane regions, those were located a world away in the Andes of South America. Here in the Balkans, though, the crop had also fared well and now represented the largest cash crop for local families.

I'd been amazed by not only the local people's agricultural skills in farming this steep landscape but also their survival skills, their ability to grow or wild harvest enough food plants in the warm months, storing them in straw-lined dirt pits, or fermenting them, for consumption throughout the long winter. Some years, the snows were so heavy and the roads were so bad that it wasn't possible to leave the villages for many months. If they didn't stock up, they'd starve.

The fermentation of wild fruits—rose hips, plums, cornelian cherries, blackberries—into fizzy beverages consumed for their perceived health benefits fascinated me. Then there were the lacto-fermented vegetables— delicious peppers, tomatoes, cabbages, cucumbers—that were pickled in brine and served with yogurt, cheese, and bread throughout the heart of winter. Locals also held in-depth knowledge of wild sources of microbes that could kick-start the fermentation of different ingredients, such as the local acidic yogurt beverage kos. Just five years before this expedition, Andrea and I coined the term *ethnozymology* to describe this domain of traditional ecological knowledge we'd noted in our prior research with the Gorani.

We continued the slow trek on curving dirt paths until I spotted it gleaming in the distance. "We're almost there," I told the team in the SUV. "That's the dome to the mosque of Borje."

Driving down the main street, we passed hand-built rock homes, their roofs thatched with a special long-stalked variety of barley that doubles as a food source. We passed by an old man with sun-weathered skin leading a donkey with firewood strapped to its back. We parked the SUV downhill from the mosque, in the shade of one of the many willow trees. Though this region is predominantly inhabited by Albanian communities, if a village in these parts of the Gora mountains is filled with willow trees, there's a good chance that it is a Gorani community. The white willow (*Salix alba*, Salicaceae) is sacred to them.

In the springtime celebration of Giorgidan (St. George's Day), the Gorani decorate the entrances to their homes and businesses with willow branches to deflect the evil eye. They also feed them to their livestock as an immunity booster. Most special, though, is the white willow's use in courtship.

When a young man is ready to propose, he walks to the river with his friends to select a special willow tree to bring back to the village. It is a night of male bonding—full of raucous singing, drinking (even though alcohol is technically forbidden by their Muslim faith), and dancing. In the end, the tree is presented to the intended bride by leaving it on the threshold of her family's home. If she accepts the proposal, she plants the tree in her father's field. If she refuses, the tree is chopped up and burned in front of the house.

In the field of anthropology, participant observation is an important methodology used to gain a deeper understanding of cultural practices; it provides a glimpse of something that of course interviews alone can't always capture. But, as a woman, observing the men search for and harvest the willow to bring back to the village would be considered taboo by locals. While my gender blocked certain avenues of investigation, it did open other doors that would be closed to my male counterparts— including greater access to young and middle-aged women to discuss topics related to women's health, childhood ailments, and childbirth.

As the team unloaded our camera gear, they looked to me for next steps. "Let's go to the bar. I can ask if Zidan is around." Zidan had participated in lengthy interviews with me during past research visits to the community; he was extremely knowledgeable about local wildlife— flora and fauna—and how the Gorani people have traditionally used these natural resources for food and health over the past century. Their cultural identity is tightly tied both to their unique language and their traditional lands. Zidan was a hunter and, like me, had only one leg.

The café-bar was a single room situated near the heart of the village, across from the mosque, housing a few tables and chairs, a small counter with a collection of soda and beer cans, and a coffee machine. Most important, it also had a bathroom—a simple closet with a hole in the ground, a bucket of water to clean yourself—which several of us needed after the long drive and our picnic lunch. As women, Susanne and I weren't supposed to enter this male domain, but I'd held many interviews in this bar and others working with Andrea and Mezahir in the past, and this had been accepted by the locals. Because we weren't local women, we weren't held to all of the same rules. Plus, being accompanied by five men on the team didn't hurt.

I reintroduced myself to the owner and bartender, Zuran, and asked about Zidan. He sent a young boy—one of many who lingered around the entrance to the place out of curiosity—to run to Zidan's home to find him.

Seven years ago, I'd sat with a number of the village children by the local well, praising them for their work on a school assignment of creating their own herbarium collections. They'd taped wildflowers and other local plants into the pages of their notebooks and had been eager to share their work with the stranger who had come from so far away to look at the plants of their village.

One boy, Shendet, was now grown up—nineteen years old—and had spent the past years between Albania and Germany, where he'd fled as a refugee and had reaped the benefits of a German education. We'd stayed in touch over Facebook and I'd messaged him the day before that we were coming. We shared a big hug when he arrived at the bar. Surprising me with his gains in English fluency from his studies, he shared his story.

Like many seeking a path to a brighter economic future, Shendet hoped he could make Germany his home, earning an income with the opportunities there; there weren't any here other than collecting wild herbs for pennies to be sold to middlemen and eventually ending up in fancy health food stores as dietary supplements across Europe and the United States. He had studied hard, hoping that his skills in four languages—Albanian, English, Gorani, and German—would serve as his ticket to a better future. But when he turned eighteen and became a legal adult, he was deported back to Albania. "They called me Albanian," he recounted as if it were a slur. "I tried to explain to them that I am not. I am *not* Albanian. I am *Gorani*." Shendet laid his hand on his chest. It was as if, with those few words, others had tried to strip him of his personhood and sense of cultural identity. As Gorani, he grew up with a language, culture, customs, and relationship to the natural world distinct from the majority Albanian culture in the country.

Shendet made me think of my children and nephew back home. I couldn't imagine their traveling to a foreign country alone at the age of thirteen to live away from family in a foster system, studying languages as if their lives depended on it, and then to have it all ripped away as soon as they hit the first blooms of adulthood. Compounding his woes, Shendet's uncle had passed and his father was ill, leaving the burden of finding money for not one but two families on his young shoulders. I ached for him. I slipped him some money to help him take care of his family through the month.

In the midst of our discussion, the room had filled with curious onlookers, mostly young men in their twenties and thirties. I told Zuran, the bartender, to put a round of drinks on me. We sat and chatted, passing the time and introducing the rest of the team to some of the traditions here. When the bill came, the conversion rate of Albanian lek to euros was so low that even after buying sodas, beers, and coffee for everyone in the packed room, I owed only twelve hundred lek, equivalent to ten euros, or twelve dollars.

Zidan arrived to a crowded room. He limped forward, and a young man got up to offer him a seat. A robust man in his late fifties, Zidan had skin that was tanned and wrinkled from the time he spent outdoors. His coal-black hair now featured streaks of silver. We exchanged greetings; he remembered me well, was happy to catch up.

"How have you been?" I asked. "Any bear this season?"

Zidan was known especially for his skills in tracking and hunting brown bears, locally known as arieu. The meat was prized for food, but the bear fat even more so for creating local traditional medicines for wounds. Though he didn't hunt to sell his catch—using it only for local consumption (for food and medicine) for his family and friends—others in the country did. The extensive hunting and trade of wildlife greatly threatened wild populations of bears, foxes, and many birds, including the eagle—the national symbol of Albania, a totem associated with freedom and heroism of their folklore.

In 2014, the Albanian government declared a total hunting ban for two years in the hope of providing wildlife time to replenish the country's populations. This came on the heels of a 2013 essay in *National Geographic* magazine, which brought attention to the declines in migratory birds in the region; there was great public pressure to take action. In 2016, the ban was renewed for another five years, through 2021. It's unclear how effective this ban is, especially in the remote regions of the country. One study confirmed that while many people are aware of the ban, there is little enforcement. In many cases, local hunters have bigger concerns, such as feeding their families, that drive these practices.

Populations of wild plants and animals alike were being threatened by people doing all they could to survive on the limited income these forests provided. Sustainable harvesting methods are needed, but as long as poverty persists here, it's difficult to imagine how such initiatives to save wild flora and fauna will succeed. Conservation can't happen in a vacuum. As long as humans are considered separate from—instead of part of—nature, such initiatives will likely be met with little success. Humans have been altering their landscapes, affecting wildlife populations, since we first walked this earth. Everything is connected.

As an ethnobotanist focused on edible orchids and their conservation, Susanne understands this all too well. She had a particular interest in salep, a group of orchids whose bulbs are wild harvested to create an array of products, from chewy ice cream to explosives—it all has to do with the unique chemical properties of the orchid tubers. A determined scientist with strawberry blond hair and pockets full of flavorful herbs she gathered on her walks, Susanne was on a mission to better understand the complex dynamics of the orchid trade, and many of the wild orchid tubers that made their way from these mountains into Turkish marketplaces, where they were sold as ingredients in a frothy, hot, sweetened beverage peddled by street vendors on winter days in Istanbul.

Zidan knew some local orchid meadows. I asked him to take us there. We couldn't all fit into our SUV, so the bar owner, happy with his good fortune in sales and enjoying the group chat, offered to drive us in an old boxy red van he owned. Along with a good number of village men, we all piled into the van and made our way out to the orchid fields, bouncing along the bumpy roads while listening to Albanian pop.

We didn't have to drive far. We stopped at a point where the dirt path turned into a wet grassy field. Zidan took off, surprisingly agile, and I tramped after him, both of us limping our way through the sodden fields and over some small streams. We crossed a barrier of shrubs and trees, and then, all of sudden, we were there.

My eyes widened. The grassy meadow was adorned with terrestrial orchids in shades of purples and whites. There were other bright blue and pink wildflowers as well. It was as if we'd popped out of one world to find ourselves in a botanical wonderland. I felt dizzy with it—though that may have also been due to the antihistamines I'd taken to combat my constant sneezing fit from the abundance of pollen.

Susanne got busy photographing and making notes on the key species locals harvested. Shendet and Zuran, both curious about her work, peppered her with questions about how she could tell the difference between one type of flower and another. Why were *these* flowers so important? Why had she traveled *so far* to see things that are normal to them? What was life like in England, where she lived, or the Netherlands, where she was completing her studies? She, in turn, asked them about the harvest and trade of wild orchids in the region. How much could be earned? Did outsiders come in to take them from these fields?

Surveying the hillside with discerning eyes, Zidan found a perfect place to sit, leaning back to enjoy the view. Even though this was a paradise he was privy to any time, he still appeared to revere its beauty. I joined him in the grass, pulled my water bottle out of my pack to rehydrate, and then peeled down my stump liner to dry my stump of the sweat that had accumulated from our walk. He pointed to the woods not far in the distance and spoke, Shendet interpreting: "That's where I stepped on a land mine and blew my leg off."

So there we sat, the one-legged hunter and me. Though, to be more accurate, there sat two one-legged hunters. Though we couldn't communicate directly with words without an interpreter, we shared an unusual bond. War had cut us both deeply. For Zidan, loss came at the hands of destructive land mines that were planted by Serb military and paramilitary forces, unmapped along 120 kilometers of Kosovo's border with northeast Albania in 1998–1999. Since the end of the war, more than two hundred incidents with land mines had occurred, mainly with villagers living in this region, some with fatal outcomes.

It turns out that, in all likelihood, my congenital leg deformity, which led to my amputation, was owing to Daddy's exposure to toxic chemicals during his deployment in Vietnam. The United States military sprayed around nineteen million gallons of defoliant chemicals onto ten thousand square miles of Vietnamese jungle and mangrove forests during Operation Ranch Hand; eleven million gallons of the spray were a dioxin chemical also known as Agent Orange. Dioxin chemicals remain intact in the environment for long periods, and exposures can occur through the skin, by breathing, or by eating foods contaminated with them. They are fat soluble, and can accumulate in the body's tissues, where it is hypothesized that they interfere with the genetic switches that control early development, when an embryo becomes a fetus. Agent Orange defoliant destroyed forests, biodiversity, and animals; it even altered the genetic makeup or expression of those exposed to it, American and Vietnamese. For me, it is possible that it disrupted the way my body developed in the earliest months of life inside my mother, where my father's mutated sperm and my mother's egg conferred the genes that directed how my body was formed.

Perhaps my bond with Zidan was about more than war and a love for nature and hunting—it was also a sense of knowing what few others seemed to recognize, knowing in our bones the consequences of destroying nature. I'd gleaned this from our long interview discussions. Wars have been fought throughout the millennia, and as humans continue to battle over the limited resources this planet has to offer, there is no end in sight. The damage to humankind doesn't end with the signing of peace treaties; nature's scars will continue to haunt us in the most unexpected ways. Zidan and I shared a respect for the awesome nature of plants.

In its very existence, ethnobotany is a signal of hope. It's where human experiences and scientific scrutiny meet—and wonderful things can happen. From my lab studies, I'd learned how very potent plants are, how poison and medicine are often two sides of the same coin. I'd also come to appreciate just how little we actually know about them. My work is driven by a core belief in the great possibilities of the natural world, and the potential that can be unlocked from these resources for the betterment of the world—that is, if we don't destroy them first.

In the Balkans, I'd found great beauty and great pain. Physical and psychological scars of violence and rape stand out in living memory. Everyone has been touched in some way. Mezahir, our radio DJ interpreter, so full of life and humor, carried a sadness in him; he was forced to watch his uncle tortured and murdered during the war. And yet, despite the scars, seen and unseen, these people have treated me with great warmth, as if I were not a distant stranger, but merely a long-absent family member.

WHILE I WAS LEADING this expedition in the Balkans, another one was underway in the boondocks of Georgia—the state, not the country. This was the first time my research group had two full-fledged expeditions underway at the same time and on different continents. This was in addition to our normal operations in the active lab, packed with summer interns. The year prior, I'd established a fruitful collaboration with partners at the Jones Center at Ichauway, located in Baker County. Consisting of twenty-nine thousand acres of longleaf pinelands and swamplands, this was one of the last remaining intact ecosystems in the southeast United States. An avid outdoorsman, Robert W. Woodruff, the former president of the Coca-Cola Company, originally purchased the property. Woodruff enjoyed bird hunting on the grounds and maintained the landscape in its natural form. Following his death in 1985, the land was dedicated as a center for environmental research. A campus with dorms and labs were built, and the center now supports a staff of more than eighty-five employees and one hundred graduate students who share the mission to understand, demonstrate, and promote excellence in natural resource management and conservation on the landscape of the southeastern coastal plain of the United States.

Working with on-site scientists, we'd undertaken an extensive survey and collection expedition targeting wild species known to have been used as wild foods and as treatments for infectious and inflammatory diseases by the Muscogee (Creek) Nation and Choctaw Nation. I felt a special tie to these pinelands—it was the same landscape that my father, grandfather, and great-grandfather before him had worked. It was the landscape where my ancestors had once hunted deer and fowl and foraged for wild edible greens and fruit. Now, it was the land where I hunted for plants whenever I could get away from Atlanta, just four hours away. It was also the land where I hunted deer to feed my family and began to teach Donato, at the age of twelve, to hunt during the fall season.

As I was working in Albania and Kosovo, the herbarium team and my graduate students in the lab were there looking for some additional species on our list. The forest understory is rich in fragrant plants like the sweet goldenrod (*Solidago odora*, Asteraceae), which smells like licorice; sassafras (*Sassafras albidum*, Lauraceae), whose safrole-rich leaves are dried and powdered into filé [or *gumbo filé*], the secret in the sauce of tasty Creole gumbo; and the more odorous American beautyberry (*Callicarpa americana*, Lamiaceae), whose leaves act as insect repellent. It's also full of rattlesnakes; I'd trained the team to be snake aware and to wear impenetrable snake gaiters over their pants when out in the field. But being so far away, I worried about them. More than anything else, they were most likely suffering from the incessant buzzing of flying gnats that swarmed around their noses, eyes, ears, and mouths.

AFTER TAKING A THREE-YEAR break from apartment maintenance to be a full-time stay-at-home dad until Giacomo began school, Marco came back into the workforce—this time as a lab technician with the research team. After years of repairing equipment, contributing to the backbreaking work of field expeditions, and assisting in the collection and processing of plants for years for free, we were finally able to secure a role for him that paid for all of his many important contributions to the lab.

While we were in the field, back in the Phytochem lab, Marco designed and custom built an open-bed fraction collector—basically a platform that could be used to capture small amounts of liquid coming off our preparative high-performance liquid chromatography system into test tubes. The trick was that by collecting the separated liquid over and over again, we were able to isolate larger quantities of pure compounds from our biologically active plant extracts. Working with limited funds to build this, he'd done as we've always done in the lab—gone through our supply of spare parts and random tools gathered from scavenging trips over the years.

This time, however, he scavenged in an unusual place—our own kids' stash of Legos. He realized that he could build the whole robotic mechanism using Lego parts and then coupled it to a Lego Mindstorms robotic controller. Then he worked with other lab members to write up the design and instructions for others to use. With less than $500 worth of Lego pieces and common supplies found in a local hardware store, he re-created a scientific tool that would cost between $10,000 to $15,000. That's my man.

Now students in the lab ran their samples over and over again on autopilot, returning to find enough pure compound in the test tubes to yield crystals. This was a tremendous breakthrough for us. We not only had tools like liquid chromatography tandem mass spectrometry to examine the chemical composition of complex botanical extracts and NMR (nuclear magnetic resonance) to identify the two-dimensional structure of pure compounds we isolated but also could now capture the exact structural conformation of the pure molecule. Using X-ray crystallography tools, we could see exactly how the compound aligns itself in three-dimensional space. This could prove exceptionally useful for future studies using computer-modeling software to visualize how different compounds can fit into protein targets, much like a key fitting into a lock.

The first crystal we nailed was one we isolated from the leaves of the American beautyberry. Beautyberry has a long history in Native American medicine. The Seminole used the root or stem bark for itchy skin, whereas the Choctaw Nation and Coushatta Tribe have used it for various gastrointestinal complaints, decocting the roots and berries for colic, stomachache, and diarrhea. The leaves are also well known for their insect-repellent effects, and scientists have successfully isolated the main compound responsible for this effect.

Though the compound emerging from our studies had been originally discovered by another research group and reported to have anticancer properties, we were the first to successfully isolate a crystal of high-enough quality to determine its 3-D structure. We also isolated and identified a compound, known as a clerodane diterpene, from the beautyberry leaves. While it had some limited antimicrobial activity on its own, we found that when paired with beta-lactam antibiotics (an antibiotic class that has a beta-lactam ring in the chemical structure) like oxacillin or meropenem, it restored the activity of the antibiotic back to the strains of drug-resistant staph bacteria (*Staphylococcus aureus*). This approach of resensitization of antibiotics is extremely attractive in the drug-discovery world because it could potentially allow us to bring some of our previously effective drugs off the shelf and back into practice to better treat infections!

We also investigated the activity of beautyberry leaf extracts on the growth of *Cutibacterium acnes*, the bacterium responsible for acne. I loved the ring to that—beautyberry leaves for beautiful skin! Importantly, it showed growth inhibition at concentrations that were not toxic to skin cells in the lab.

We'd had a false lead the year before on a mushroom extract that obliterated acne bacteria in a petri dish—even in its super-resistant-biofilm life stage. But when we tested the mushroom extract on a dish of skin cells in the lab, it wiped those out as well! Think: *total face peel*. While there are many compounds from nature that can kill microbes, the trick is finding the ones that do so *without* also killing the human cells. It takes a lot of patience and diligent searching to find those. Not everything that is natural is safe.

Once we had these early successes with beautyberry, others soon followed. Our work on the chemistry and antivirulence properties of extracts from the Brazilian peppertree led to the successful isolation of three active compounds from the peppertree fruits and to determining their 3-D structure by X-ray crystallography in addition to NMR. These compounds showed promising activity not only in test-tube studies but also in skin infections with live animals.

I couldn't wait to dig into the chemistry and test the antimicrobial activities of the houseleek (*Sempervivum* species) that we'd scrambled up cliffsides in Kosovo to collect. Its botanical family, the Crassulaceae, had caught my attention from some interesting findings we'd already uncovered from another genus in the group: *Kalanchoe*. Originating in Madagascar and southeast Africa, the fleshy tropical species *Kalanchoe mortagei* and *Kalanchoe fedtschenkoi* share a long tradition of utility in traditional medicine. Due to their ability to self-propagate from plantlets produced on margins of their leaves, these species can now be found globally across the tropics, where they are frequently called "miracle leaf" for their many different medical applications.

Our examination of the species revealed that, of the two, *Kalanchoe fedtschenkoi* exhibited the most potent antibacterial effects, inhibiting the growth of multidrug-resistant gram-negative species (*Acinetobacter baumannii* and *Pseudomonas aeruginosa*) as well as the gram-positive pathogen *Staphylococcus aureus*, while exhibiting low toxicity to human skin cells. Our next steps would be to isolate the active chemicals of this species, and I was excited to see if the Kosovar *Sempervivum* species would yield any similar chemicals with antibacterial activities in lab tests. While we still had much work to do on this and other projects, we were setting a new bar for research quality and rigor in the field. I couldn't have been prouder.

CHAPTER 12

Cassandra's Curse

The pandemic reminds us there are no differences or borders
between those who suffer. We are all frail, all equal, all precious.
May we be profoundly shaken: Now is the time to eliminate
inequalities and heal the injustice undermining the health of the
entire human family!

POPE FRANCIS, VIA TWITTER, 2020

In the DC hotel room, I paced back and forth from the edge of the bed
to the small seating area, where an L-shaped couch faced a sleek wall-
mounted flat-screen TV. Cable news hummed on low volume. It was
mid-February of 2020. I'd spent the weekend bundled up with a heavy
coat while cheering along the sidelines of a soccer field where Bella's
team played a rival group in South Carolina. We made the drive back to
Atlanta just in time for my late-night flight up north. Over the past few
months, I found myself on a plane either crisscrossing the Atlantic or to
the West Coast of the United States and back. The schedule of speaking
events at universities, scientific conferences, and job interviews was
both exhilarating and exhausting. I'd just reached the end of a long
day on Capitol Hill, and my leg ached from walking in the cold wintry
air—hours racing from one building to another with a small team of

scientists, survivors, physicians, and public health officials from around the country.

We'd been brought together for an advocacy day event by the Pew Charitable Trusts to meet with members of congress and senators (or their health and science advisers) about the urgent threat of antibiotic resistance. Armed with bright blue bags with STOP SUPERBUGS! stamped across them, we spent hours in meetings explaining the need for more congressional funding to support antibiotic discovery and development research. We emphasized the importance of public health initiatives to stop the spread of multidrug-resistant infections in our hospitals and communities. The meetings went well, but who's to know what politicians will do in the end?

We pleaded our case honestly and clearly. Our call was simple—if our government didn't make antibiotic resistance a priority, it would one day cripple the entire medical system. Elective surgeries would be life-or-death decisions; so, too, would everything from childbirth to cancer therapy. In 2019, the Centers for Disease Control ranked sixteen bacteria and two fungi into categories based on three threat levels they posed to humankind: urgent, serious, and concerning. My lab was actively studying seven of these deadly pathogens in the urgent and serious threat categories, looking for new chemicals from nature to kill them, restore the activity of existing antibiotics with combination therapies, or knock out the virulence pathways that ravaged tissues.

There was good reason for these politicians to be concerned. The scourge of tuberculosis, an airborne threat that has haunted humans for at least nine thousand years, had evolved not just multidrug resistance, but also extensive drug resistance; some strains were now resistant to two of four first-line antibiotics, and another also to at least one of three second-line antibiotics. Now, total drug-resistant TB was emerging. *Mycobacterium tuberculosis*, the bacterium that causes TB, was discovered by

Dr. Robert Koch in 1882; back then, it killed one out of every seven people in the United States and Europe. The past is a clear window to the future—this is where we are heading.

We were running out of options. Progress on the discovery of new antibiotics is slow—partly due to the rate at which the bacteria grow. While fast-growing bacteria like *E. coli* have a doubling time (the time it takes to double the number of cells in a culture) of just twenty minutes, enabling scientists to set up an experiment and get results the very next day, the doubling time for *M. tuberculosis* is twenty-four hours or more—meaning scientists must wait a minimum of two weeks to glean the results of a single experiment.

It wasn't just the airborne pathogens that were a cause of worry. Untreatable strains of "super gonorrhea" caused by the spread of *Neisseria gonorrhoeae* are on the rise. Gonorrhea can infect women and men in the throat, rectum, and genitals. When left untreated, it causes pelvic inflammatory disease in women (symptoms include long-term pain, infertility, and ectopic pregnancy) and pain in the testicles and sterility in men. Sexually active young people from fifteen to twenty-four are most commonly afflicted, and women can also pass it to their newborns during childbirth. In 2018, there were more than 550,000 cases of gonorrhea in the United States. Left unchecked, it can spread in the blood and joints, and even kill. While the drug-discovery pipeline for resistant gonorrhea is quite sparse, there are a couple of promising candidates under evaluation in clinical trials: zoliflodacin and gepotidacin. Even if these pass muster for drug approval, the pipeline needs to be refilled. History has taught us that resistance is inevitable.

I paused to gaze at the TV. For a few days now, I'd been keeping an eye on a newly emergent respiratory virus from Wuhan, China. News outlets made it sound as if the coronavirus were happening on another planet, but I knew that distance doesn't matter when a virus can spread quickly from person to person. Based on the early reports, the virus was passed easily via respiratory routes and possibly by contamination of surfaces. It had reached northern Italy: hospitalization rates were skyrocketing, especially among the large elderly populations. The cases overwhelmed the ill-equipped hospitals. I worried about Marco's family in the south of the country, praying that their small village would be spared.

Along with the world's brain trust on antibiotic discovery and development—government, academia, military, private industry scientists, and physicians—I was scheduled to convene in a few days at a hotel in Barga, located an hour or so inland from Pisa, near Lucca, Italy. The theme was "Challenges and Innovative Approaches to Discover and Develop New Antibacterial Agents." I was looking forward to not only presenting our latest discovery of a compound that restores the activity of beta-lactam antibiotics in MRSA infections, but also catching up with old friends and brainstorming new creative paths forward.

My phone buzzed—a new text from Dr. Ryan Cirz, one of my colleagues who had succeeded in bringing a new antibiotic (plazomicin for multidrug-resistant, gram-negative bacterial infections) to market through FDA approval, only to watch his company, Achaogen, fold under the dysfunctional economic model that antibiotics are stuck in.

While the costs of making drugs that people take for chronic conditions—such as depression, heart disease, and high blood pressure—can be recouped over long periods of use, the story is different for antibiotics. When used properly, antibiotics should be administered over only a short period (a couple of weeks at most), limiting the time to recoup those expenses. And then there are other public health concerns to consider—for the public good, doctors use the older drugs as long as they still work, keeping the newer drugs in reserve for situations (new, emerging, resistant strains) that the older drugs can't tackle. This is because unlike beta-blockers and statins, antibiotics lose their efficacy over time as new lines of resistance emerge. So even if a company brings a safe and effective antibiotic through the rigors of FDA approval to reach the market, the company is faced with trying to recoup investments on a drug that is used for only a short period per patient, is held in reserve for only the worst resistant infections, and has a limited life span itself before it stops working because of bacteria becoming resistant to it as well!

You may be wondering why anyone would dedicate their life to research that seems destined to fail. Although the economic model is broken, scientists aren't giving up. The risks to humanity are too great to quit; we know that the problem of antibiotic resistance isn't going away and that we're standing on the precipice of a post-antibiotic era. We need to be ready for it.

Ryan texted that the meeting was still "on," although many of us were becoming uneasy. The infection map had not yet hit that particular town, but it was creeping. We needed this meeting, but at what cost?

Ryan's text explained that two more plenary speakers had dropped out, as well as the entire UK government scientist contingent. I rang a close collaborator, Dr. Dan Zurawski, chief of pathogenesis and virulence in the Wound Infections Department at Walter Reed Army Research Institute, who'd been working with my lab on the discovery of new compounds for deadly carbapenem-resistant *Acinetobacter baumannii* infections. Surely if there were news on threat levels, he'd be most informed. His trip was still "on" for the moment, he said, but it could change at any time.

All this uncertainty worried me. There were no treatments or vaccines, and the case numbers of severe infections and deaths were rising quickly.

In addition to these larger-scale issues, Marco and I had scheduled the closing on a house for the day I was set to return from Italy. After a year of soul searching and job hunting across the United States and Europe, I'd made the decision to stay at Emory under a renegotiated contract that gave me more support and stability. After burning through money on apartments and home rentals during my education, training, and early career, at the ages of forty-two and forty-nine, we were finally settling down.

If I became sick or needed to be quarantined while out of the country, it would be disastrous for the family. Because the conference organizers remained in wait-and-see mode, I made my decision to head home and stay there.

When I arrived home, Marco and I stocked up on nonperishable food, cleaning supplies, and vegetable seeds, and followed the disaster readiness guidance. All those years of living through Florida hurricane seasons had prepared us well. Once the rest of the city caught wind of the virus's spread and the lockdowns that would surely come, there would be a run on everything from groceries to gasoline.

I told as many of my friends as I could to prepare for a lockdown and to stock up on enough food and supplies for a month of isolation. Some listened, others deemed me crazy. How could a virus affecting people across the ocean affect us? Two words: air travel. We were only one direct flight away from China, and the virus was likely already circulating undetected in the air from person to person in the United States.

Three weeks later, on March 11, 2020, the World Health Organization (WHO) announced that COVID-19 was officially a pandemic. Before the schools ever announced closures, I told my children to pack up their lockers and gather their personal items—they wouldn't be going back to school for the year. I instructed my research team to cycle down experiments and shut down the lab a week before the university implemented shutdown procedures. With college spring breakers returning and the high likelihood of some being infected or carriers, I didn't want to risk having any members of the team on campus then. We packed our bacteria away safely in deep freezer storage and sterilized the lab surfaces. They took home computers and digital copies of their lab note-books and established work-from-home stations. We got out before the rush in a relatively calm and organized way. They felt safe and secure at home, as did I. Because of that security, we were able to get back to the business of science quickly, working as a team on writing several big sci-entific papers and grants we'd been putting off.

While my team and I focused our efforts on finalizing two large re-view papers on plant-derived antibiotics from the safety of home, I watched in awe as my healthcare colleagues suited up to battle this mi-croscopic foe in the hospitals; epidemiologists tracked cases as infection rates peaked from country to country and state to state in the United States.

We'd made the move into our new home just before the pandemic hit. Marco juggled his databasing lab work with other construction projects for our home, building custom bookcases for the home office as well as raised beds for our backyard vegetable garden. My work schedule at home became the busiest it had ever been. Our sixteen-year-old nephew, Trevor, joined us to do remote schooling in the company of his cousins. I took to waking up at five a.m. every day to get additional work time be-fore the kids woke up for virtual school. I managed the lab group re-motely, relying on group chats, regular video calls, and massive to-do lists. I struggled to take care of my own writing and revising work—grants, papers, and book drafts awaited.

And then there was all the at-home work. Whatever time I saved by not commuting to work was swiftly taken up by new responsibilities I shared with Marco as a full-time chef, K–12 teacher, and maid, all in addition to my professor job. When Marco returned to work in the lab in June, it left me juggling many of these tasks—and especially the kids' education—on my own during the workdays. Giacomo, our youngest at seven, needed the most help. Our saving grace was a poster-board-size daily schedule I'd written out for the family and taped to the wall—it covered everything from chores to studies to snacks and mealtimes. The kids had to step up in assisting with tasks, and they got onto the cooking and cleaning rotations.

Hand sanitizer was in short supply for my physician colleagues, so Marco and I made fifteen gallons of hand sanitizer in the lab with our supplies of isopropyl alcohol, hydrogen peroxide, and glycerol—mixing them following the ratios provided in the WHO's recipe. We delivered individual spray bottles of the sterilizing hand rub to my colleagues and big clinical research teams working on antibody diagnostic trials. We distributed jelly jars full of bleach, hand sanitizer, and home-sewn masks to members of my research team, leaving bags of goodies at their doors so they could safely go out for groceries.

Beyond these supply deliveries, we tried not to leave home for those first two months. With a household full of hungry mouths to feed, we came up with creative ways to eat through our food stockpile. We organized two fifty-pound bags of potatoes and onions, boxes of durable cabbage and acorn squash, packs of assorted dried beans, canned foods, boxed milk, and a freezer full of venison from my hunting season the previous fall. Our garden was way behind schedule, and the most we could harvest in those early days were fresh herbs and a few tomatoes, but when the summer arrived, we delighted in the bounty of squash, cucumbers, tomatoes, herbs, peas, okra, and beans that we'd planted.

We were the lucky ones. While faced with the stressors of finding a new rhythm at home, families across the globe were living the nightmares of bringing sick relatives to the hospital, only to be separated from them at the door, never to be seen again. Patients were dying in hospital rooms and hallways, the last words spoken to their families over a nurse's or doctor's phone or a video call on a tablet, never to feel their comforting touch again. And the numbers of cases and deaths kept rising.

. . .

Just three years before, we'd had our own brush with death in the family. The trauma of the experience was still fresh in our minds as we took in the chaos that the pandemic brought with it.

I'd been in a meeting when I received Marco's text.

"I'm going to the ER. I don't feel good," he wrote.

I typed back a rapid response to stay at home and that I'd come pick him up right away.

"No, no—don't worry. It's probably nothing. I can drive. Just pick up the kids after school."

He'd been feeling off for the past day or two and had taken a sick day to rest up at home. He thought he might have overdone some yardwork, straining his muscles. His fingers felt odd, tingly. He had difficulty picking up Giacomo at bedtime the night before. He thought an X-ray and a quick exam could clear up the issue.

Each time I called or texted to check on him the rest of the afternoon, he insisted he was fine and that there was no rush for me to join him at the hospital. He'd be home soon. After picking up the kids from school and arranging for a neighbor to babysit, I went to the hospital to check on him. He'd been there for hours already. Maybe they were backed up. The ER can be slow.

When I arrived, he was gone from his exam room. The nurse explained that he was getting a CT scan. He looked pale and wan when they rolled him back into the room in his hospital bed. "What happened?" I asked softly, gently brushing his hair back over his forehead in a caress, trying to understand the full story.

Marco had slept until noon, he told me, but when he got up, he could barely walk. The master bedroom in our rental house at the time was downstairs one level from the main floor. He dragged himself up by gripping the handrail. When he managed to reach the kitchen and poured himself a cup of coffee, it tasted funny. He called his mom and she'd persuaded him to go to the hospital. "You drove in that condition?" I asked, confused and upset. "You should have told me, I wanted to come right away."

A neurology resident came into the room. I'd already begun to panic, but I worked to keep outwardly calm. It wouldn't do Marco any good to see me break down. What could cause this kind of rapid neurological degeneration in a man in his forties who was in peak fitness and health? Multiple sclerosis? A stroke? Poison? A rapidly spreading tumor? What was the threat?

The doctor helped Marco sit up in bed and ran a sharp object along the soles of his feet. Marco couldn't feel a thing. He poked Marco with a sharp tool, jabbing him, but not breaking skin. No sensation in his hands and forearms, or his lower legs or feet. The doctor explained that the CT scan of his brain looked fine, but they needed to run more tests. Marco would need to be admitted.

I called our friends Matt and Jen—also parents of three—whom I'd been close with since college. As I explained our crisis, they tried to reassure me that all would be okay. Matt would bring dinner to Granny and the kids, and would relieve the babysitter and stay with them until I could get home later that night.

The next morning was a Saturday, and the kids were eager to share how cool it was that Matt brought them a special treat of cheeseburgers and ice cream! But something was off, and, as ever, our most observant child, ten-year-old Bella, was more suspicious than the others. "Where's Papa?" she asked me, her eyes inquisitive and piercing. Donato, now in his first year of middle school, chimed in with his disappointment. They were supposed to practice soccer drills together that day.

What could I tell them? I couldn't share my fears. We didn't know what was wrong or how serious this was.

"You know how I have this big research project due soon? Well, we had to work late last night in the lab and Papa got up early to go back this morning before you woke up," I explained, fearing they'd see through my lies.

Donato retorted, "Mamma, don't you have other people working for you? You should give Papa the weekends off." I reassured him that I would and explained that Sahil and his daughter, Larkin, whom they loved to play with, were coming over to spend the day with them while I finished up work with their dad in the lab.

I reached the hospital to find that Marco hadn't slept well, his back ached with shooting pains that plagued him all night, the tingly feeling worsening in his limbs. In addition to multiple blood draws, they'd also done a spinal tap to examine his spinal fluid. The poor man had been poked and prodded continuously, but it was necessary; we needed answers.

When the results came back later that afternoon, the attending neurologist explained that they were confident in the diagnosis. It was Guillain-Barré syndrome. I'd heard of it before, but knew very little about what caused it or what to expect. The doctor explained that it was a very rare disease: only twenty thousand people get it in the United States each year, or between one and two people per every one hundred thousand. She asked Marco about any recent illnesses he'd experienced. In fact, just three weeks before, we made our annual road trip to Florida to spend time with family over the Thanksgiving holiday. We stopped at a gas station to fuel up and he grabbed a chicken sandwich to eat from the food counter. He was the only one in the family who ate it, and he came down with an awful case of food poisoning—vomiting, diarrhea— that knocked him out of commission for two days.

It turns out that *Campylobacter jejuni*, a gram-negative bacteria and common cause of food poisoning—especially in undercooked poultry—can also trigger your immune system to go into hyperdrive and start attacking the myelin sheathing that coats the long nerve fibers running through the body. Those myelin sheaths act like the insulation coatings for the wires in electrical systems—without the insulation, the signal fails; his nerves weren't able to transmit their signals to his brain. That was why he experienced the ascending weakness and tingling up his arms and legs. He was gradually becoming paralyzed . . . and all because of a gas-station snack!

I sat by his bed, holding his hand. I didn't know how much he could even feel my touch, but I wanted to give what emotional strength I could.

"You should know that the good news is that while this is a rare disease," the doctor said, "we actually treat many patients from the southeast United States here at Emory Hospital. You're in good hands. Most patients are able to walk again after physical therapy within six months, but it may get worse before things improve. If the paralysis begins to affect your ability to breathe, you'll need to be placed on a ventilator in the ICU."

I squeezed his hand and tried to blink back the tears. His body trembled in fear and pain. His condition had rapidly worsened in the last twenty-four hours. An alarming 30 percent of Guillain-Barré patients require mechanical ventilation. If it came to that, I knew the risks that the doctor hadn't elaborated on: once he had that breathing tube in his throat, he could develop ventilator-associated pneumonia (VAP). Between 9 and 27 percent of patients on mechanical ventilation develop VAP and require extensive antibiotic therapy, and 9–13 percent of these patients die, men being in the higher-risk category. Even a 10 percent chance of death was too high for my mind to wrap around!

My knowledge of the antibiotic-resistant pathogens that attacked patients in exactly his condition formed the basis of the living nightmares in my mind. I thought of the antibiotic-resistant ESKAPE bacteria we battled in the lab: *Staphylococcus aureus*, *Pseudomonas aeruginosa*, *Klebsiella pneumoniae*, and *Acinetobacter* species were all common causes of VAP, with staph and *Acinetobacter* leading the pack. These were aggressive pathogens that could destroy tissues and evade any antibiotic hellfire the doctors threw at them.

They could kill Marco.

Now, I had to tell the kids. I needed to bring them here before he was covered in tubes and they'd be blocked from visiting him in the ICU. He could be here for days or weeks; there was no way to tell. Marco agreed, we had to do it that day. I drove home and gathered the kids. Sahil had worn them out playing games in the yard. They'd had a fun-filled day. I confessed to my lies about their dad being in the lab, but I wouldn't give them the full truth, not yet. I told them that Marco wasn't feeling well and that his doctor wanted him to sleep in the hospital to get some special medicine.

When we arrived at the hospital, the four of us held hands. Like a mother hen with her chicks lined up from oldest to youngest, I led them down the busy hospital halls and took the elevator up to the neurology floor. Matt had come earlier to visit with Marco, and he helped him sit up in bed. The kids ran up to him, eager to cover him in hugs and kisses. Giacomo offered his favorite stuffed animal to keep Marco company while he took his medicine. They brought a family photo to keep at his bedside. They chattered about their day, and he nodded with close attention. He was tiring quickly, though, and I rounded up our brood before it sapped too much of his waning energy with a promise to return soon.

As we rode down the elevator, Donato whispered in my ear to prevent Bella and Giacomo from hearing, "Ma, I think Papa was lying." His words surprised me. I turned to look closely at his face, placing my arm around his shoulders. His deep brown eyes shimmered with tears as he continued, "He's sicker than he showed. Papa's just trying to be brave for us."

A week later, after multiple rounds of intravenous immunoglobulin (also known as IVIG) composed of antibodies gleaned from donated blood, Marco was ready to be discharged. We'd been extremely fortunate that he hadn't needed the ventilator. As soon as he was able, he was up walking the hospital halls with the aid of a nurse or leaning on his IV pole. When he returned home, he was still weak and numb. At dinner, he struggled to keep liquids in his mouth. The kids found this hilarious. The comic relief of seeing their dad dribbling his juice like Giacomo had done as a baby was too funny to hold in their giggles. Marco and I smiled at each other over their laughter. We all needed to reclaim some joy after this experience.

Ever determined, as each week passed, Marco pushed himself with his physical therapy exercises, walked the halls at home, willing his body to heal. Two months later, he'd returned to a semblance of normal. While the doctors explained he'd never fully recover, he had reached a solid 90–95 percent of his function, and, to his delight, they cleared him for return to his regular work schedule.

A FEW WEEKS INTO THE PANDEMIC, I received an unusual friend request on Facebook. It was from Don Antonio. I was so surprised and delighted that he'd made the leap onto social media! It had been a few years since we'd spoken. I had hoped to travel to Peru the previous summer with my family for a visit—marking twenty years since my first trip to the Amazon. The timing just seemed right, and the kids were old enough to appreciate and remember the incredible experience of traveling in the rain forest. When other work had come up—speaking engagements and a trip to the Balkans with *National Geographic* and my Kosovar collaborators—I'd put it off for another year.

I opened the message, slowly translating Spanish to English in my head as I read. I was rusty. I quickly realized two things: Don Antonio had not written this message, and I would never see him again. It was from his daughter. He had passed away. He was in his eighties, and while I shouldn't have been so shocked that this could happen at his age, I was. He was the constant for me in the rain forest—as durable as the flow of those vast muddy waters crossing from one side of the continent to the other.

She sent me photos of Don Antonio and me together, as well as my college graduation picture I'd gifted him to remember me. She wrote that he spoke of me and kept those photos on display at home. I sat at my computer staring listlessly at the screen, weeping with deep, ugly sobs that left me gasping for air. My heart ached as if I'd lost a piece of myself in the forest. I'd lost a friend, a teacher, and a second father. Our love and respect for each other had been faithful and enduring. Regret cut at me like a dull blade. I'd made the wrong choice—I should have gone last year. The chance was gone.

A few weeks later, at midnight, the phone rang. Groggy from just having fallen asleep, I didn't pick up initially. I glanced at my dresser and saw that the call was from my friend Janelle in Canada. She must have called by accident this late. But then it rang again, and I answered. Her voice was shaky. "Cassy, I'm sorry for calling so late. I don't know how to say this, but I knew you'd want to know. Justin is dead."

Her words didn't compute. Justin was a dear friend and ethnobiology colleague to us both. During our years in Arkansas, he'd made Marco and me feel so welcome. We'd spent days together exploring the mountains and woods with him in the Ozarks. He was the same age as Marco, their birthdays just one month apart. It didn't make sense. He'd died suddenly of a heart attack and had been found at home earlier that day.

I cried for days. How much more tragedy could I take? I was at a tipping point, close to the cliff of depression and looking down, about to lose my footing. Humans are meant to grieve together—to hug and hold one another. To seek solace in the good memories and find joy in the retelling of funny stories from years past. COVID-19 had stolen that from me. It had stolen that from millions of people grieving across the globe as the infection spread and deaths continued to rise. Whether dying from COVID-19 or other causes, the rituals of everything from childbirth to funerals were disrupted by this horrific pandemic. When it came to death, closure and acceptance were hard to grasp from afar in sterile isolation.

I tried to hide my despair from Marco and the kids, but they could sense my sadness and fear. It permeated the very air I breathed. I could see it in their eyes and feel it in their sweet, unsolicited hugs.

Spring shifted into summer. I lay in the grass in the front yard of our suburban home, gazing upward at the branches of the white oaks and tulip poplars that stood sentry. Freshly cut grass poked at my skin. Ants tickled as they crawled across my arms and leg. I took deep breaths and willed myself to reconnect with nature, yearning to heal my heartache, for the roots of those magnificent trees to shoot up out of the ground and envelop my body in an embrace. I dreamt of the days in the Amazon forest with Don Antonio, wishing we could be there together in that moment. Memories of his healing ceremony fluttered through my mind— my thoughts drifted to his prayers to the forest spirits, the lulling sounds of his whistle, and the rhythmic touch of the healing shakapa bundle brushing against my body. I imagined myself a part of the ecosystem, a creature of the earth, rooted in fertile dark soils that could hold me steady through the storm.

As we neared the end of the summer, the rising use of antibiotics to treat secondary infections in COVID-19 patients, especially those on mechanical ventilators, began to greatly concern me. This could drive the rise and spread of even more strains of antibiotic-resistant bacteria in hospital wards. My team's research on the search for new antibiotics had never felt more urgent, and we pushed ourselves to regain a semblance of our old work pace back in the lab.

I also wanted to do something to fight the virus directly. There was just one problem: I was not a virologist. Though all under the umbrella of "microbiology," the life cycles and behaviors of bacteria and fungi differ greatly from those of viruses. I had much to study. I read the papers coming out on SARS-CoV-2, the virus responsible for the COVID-19 disease, to learn more about it and meditate on ways my research team could contribute. I wanted to test my plant extract library against it. Perhaps there was something in nature that could fight this.

SCIENTIFIC RESEARCH LABORATORIES have different biosafety clearances. These are known as biosafety levels—designated with the abbreviation BSL followed by a number. The number indicates the level of safety precautions needed for the containment of dangerous biological agents.

Most university microbiology research labs are BSL-2, whereas teaching laboratories are usually BSL-1. The BSL-3 labs, however, are much less common and require much higher precautions because the germs in these facilities can cause serious—and even fatal—disease via inhalation routes. Scientists work on pathogens like tuberculosis and plague bacteria, as well as respiratory viruses like those that cause yellow fever, Rift Valley fever, SARS, MERS, and now COVID-19 in BSL-3 labs. The labs require specialized engineering controls in the building to prevent any dangerous germs from escaping. There are only around two hundred BSL-3 labs in the United States, and the limited numbers and access to such facilities created a bottleneck on research on SARS-CoV-2.

At least the virus wasn't BSL-4 rated. There are only thirteen operational or planned BSL-4 labs in the entire United States. This is where scientists work on easily transmissible pathogens like Ebola virus, Marburg virus, and Lassa virus, all of which cause fatal disease.

Screening thousands of compounds against a BSL-3 virus such as SARS-CoV-2 in drug-discovery studies just isn't feasible due to both the cost of and access to the specialized spaces and scientists with the rigorous training to handle these dangerous pathogens. This is why many scientists started using in silico screening techniques early on. *In silico* is just a fancy term for computer modeling. Once the structures of some of the key building blocks of the virus were known, scientists ran computer simulations to see whether any compounds fit the targets, much like a key into a lock. These computer libraries are based on compounds known to humankind, but they don't cover the full diversity of life's chemistry.

This is where plants come in, with each tissue of each species offering hundreds to thousands of unique molecules to examine. The limitation, of course, is that because we haven't explored the vast majority of plant life for medicinal compounds, we had no idea what compounds could even be present. A computer model wouldn't work in this scenario. I needed to test the plant extracts against the virus with direct experiments, not simulations. But first I needed a more accessible BSL-2-safe option to narrow down the field. And so I searched and read and waited for our moment to leap into the fray.

BELLA AND GIACOMO RAN into the solarium, where I stood in my home office. "You're screaming, Mamma. Why are you screaming?" Concern laced Bella's question.

I lifted Giacomo in my arms and spun him around, laughing with a big goofy smile as we brushed against the pots of houseplants and palms that filled my office.

"The money came through!" I told them. "We can start the COVID project."

After keeping up an insane pace of work from home, submitting five grants over the past two months to start our work on COVID-19, I'd received one rejection and no news on the others. Amazingly, even during just two months of my team's writing the grants, the numbers of cases and deaths across the United States and the world continued to climb, and I had to update the numbers in the significance section of the grants with each submission. At last, one grant had finally come through!

It would pay for the salaries of key team members and buy the costly supplies to successfully test my extensive chemical library of plant extracts for potential inhibitors of SARS-CoV-2 viral entry. Just a few small vials of the pseudotyped virus we needed to use as a tool cost us $16,000, and that was with a discount! This was the launching pad I needed to ramp up our antiviral-discovery work in the lab. If a safe and effective botanical ingredient with protective activity against COVID-19 exists, the likelihood of it being in my library was pretty high.

I built the QNPL through collecting wild plants and subjecting them to the painstaking processes of botanical identification and authentication, chemical extraction, and preparation for biological assays. I now had more than 2,000 extracts from over 650 species used in traditional medicine for infectious and inflammatory diseases. Moreover, the collection continued to grow as I sourced new species from the commercial sector to include all the top herbs of commerce—especially those used as dietary supplement ingredients.

By early August, data on the potential activity of herbal remedies against COVID-19 was still nonexistent. However, many in the United

States continued to consume dietary supplements ranging from elderberry syrup to multivitamins to boost their immune system. I wanted to
test the top herbs of commerce first, my rationale being that if a plant
ingredient that was already in the global production chain had some
activity, we needed to identify those first to most rapidly translate those
to the scientific community and the broader public.

THE CELLS ARE GROWING well, I just split another batch yesterday,"
Cate, one of my three graduate students in the pharmacology program,
told me over one of our regular video calls. "Micah and I are planning
to run another treatment batch tomorrow. We should have enough cells
for two plates."

We'd found a BSL-2 level SARS-CoV-2 (2019 coronavirus) model for
sale from a scientific supply company based in California to use for our
COVID-19 drug-discovery screen. While bacteria are typically easy to
grow—simply add cells to a nutrient-rich media and store at room tem-
perature to achieve growth—viruses are more complicated. First, viruses aren't technically alive, and they need a host cell to enter and
undergo replication to create more copies of themselves. Unlike bacteria,
human cell lines can be slow growing and require particular environments in incubators, supplemented with carbon dioxide and the right
amount of humidity in the hot growth chambers. The good news was
that my lab was already equipped for work on human cells as well as
bacterial, fungal, and viral pathogens.

When it came to the SARS-CoV-2 virus, we couldn't handle a fully
intact virus in my BSL-2 lab, but we could work with parts of the virus
engineered for safer testing. We were working with a pseudotyped virus
made of the pieces responsible for viral entry into the cell (the spike protein) and linking it to a luminescent (glow-in-the-dark) reporter. The basic

idea is that when the virus particles enter the cell—in this case, a human kidney cell that expresses the Angiotensin-Converting Enzyme 2 (ACE2) receptor also found in lung cells—it triggers a chain reaction that results in the release of light. In other words, if the virus enters the cell, then we'd see a glow-in-the-dark response. No virus in the cell meant no light works. I have a machine in the lab that can measure signals like optical density, fluorescence, and luminescence in microtiter plates full of small experimental reactions happening in fifty microliters—the size of a raindrop.

If we could stop the viral entry step, then this could perhaps slow the spread of the virus from cell to cell once in the body. It could also potentially provide a path to developing a prophylactic, or preventive medicine, for people at high risk of exposure to the virus, such as healthcare professionals. We planned to screen the entire collection of extracts first at a single dose, and then follow up on the most promising finds to test a range of doses and examine safety with human cells.

When not loading plates, Marco kept up the production line of creating new extracts of plants used as dietary supplement ingredients and the plants that people used in traditional medicine and shipped to us to test. Once we had some viable leads, Gina and James, the PhD chemists on the team, worked on analyzing the chemical makeup of the active extracts.

We were working as fast as possible, but with safety limits on numbers of people per square foot in the university labs, we were restricted to three people in each lab at any time, requiring us to work in shifts at odd hours. We'd achieved much of our progress in antibiotic discovery research through training and working with undergraduate research interns and visiting scholars from other countries who came with their own funding. In an average year, I had as many as thirty people on the team. With students learning remotely and a hiring freeze in place, I was down to a lean core staff of eight. It would have to be enough for now.

Despite the setbacks, I was determined to push forward. I had a responsibility to explore the full potential of nature to combat infections. I had hope, faith, and grit on my side, as well as a one-of-a-kind library of compounds derived from plants used in medicine for centuries to treat infectious and inflammatory diseases. It wasn't a question of if we would find some hits, but rather how many, and if any would stand up to the next round of testing in live infection models.

ONE DAY IN DECEMBER at our team's weekly Zoom group meeting, I announced, "Twenty for 2020. You thought it was impossible, but we did it! Congratulations, you all should be incredibly proud of yourselves!"

A year before, I told them that I wanted to do "something big" to commemorate the entry into a new decade—the eighth year of operations for the research group.

James had quipped, "Maybe we can get twenty pubs for 2020."

He was joking, but we'd made it a reality. He wasn't talking about beer-fueled pub crawls but rather twenty scientific publications. The time away from experiments during the early months of the pandemic gave the team the critical time needed to work collaboratively on wrapping up a series of original research and review articles as well as new patents. The hard work paid off. We succeeded in publishing twenty new articles on our work toward the discovery of plant sources of compounds to combat drug-resistant infections, and we had a number of promising leads on projects to pursue. This included species-specific growth inhibitors from Brazilian peppertree and Sicilian sumac that could take a more selective approach to targeting pathogens (bad bacteria) while leaving commensals (good bacteria) intact. There were specialized compounds from the American beautyberry that restored the activity of beta-lactam antibiotics in treating MRSA. From chestnut and peppertree, we identified new molecules that could effectively switch off the communications system that staph use to release toxins and cause massive tissue damage, reducing the harm they could inflict. We found new leads on plants that could help in the fight against periodontal disease, and continued work on others that could prevent the formation of biofilms and enhance antibiotic efficacy in the removal of biofilms from medical devices. Critically, we'd also found solid scientific evidence that pointed toward the efficacy of many previously neglected traditional remedies, which would've once been dismissed as folklore by modern science. We were building a case for the amazing potential that plant medicines hold for the development of future medical innovations to improve quality of life and longevity in the face of infectious disease.

COVID-19 HAS LAID BARE the threat of infectious diseases that lack effective treatment, which is also a lack of preparedness. Collectively, humans across the planet experienced a massive shift in the ways we live, eat, socialize, travel, work, and learn. The threat of grave illness and even death was there for all of us. No one was safe.

Science takes time, money, access to the right resources, and team members trained in the skills needed to run complicated experiments. Leaps in progress don't happen overnight; they take years of investment in training and research, laying the foundation in basic insights in order to be ready for translation and innovation. Congress needs to act now, and decisively, to invest in research and training of scientists and physicians to fight the war against antimicrobial-resistant superbugs. Funding rates for graduate student PhD training grants from the National Institute of Allergy and Infectious Diseases (NIAID) of the NIH are extremely low, and it has reached the point where I wonder whether it's even worth the effort for my students to apply to a fund with so little money to go around. The lack of funding priority for infectious diseases doesn't just affect scientists. Incredibly, physicians who specialize in infectious diseases are among the lowest paid of any specialty field of medicine, earning less than half the annual incomes of gastroenterologists or cardiologists (an annual salary difference of $200,000 or more). The response to COVID-19, with an investment of $18 billion into the US government's Operation Warp Speed, has illustrated the power of funds to push innovations out. COVID-19 has been our proving grounds; let's not wait until the next Operation Warp Speed is needed but instead seriously dedicate ourselves to the work now, before it is too late.

In Greek mythology, Cassandra, one of the princesses of Troy, was gifted with great beauty and the ability to foresee the future. Cassandra's curse was that no one believed her predictions, and Troy was destroyed during the Trojan War.

But we don't need the gift of foresight to predict what is to come. Infectious diseases have plagued humanity since the time of our origins. To be prepared for the next threat, we must invest in the future.

The economic model for antibiotics requires a major shift. In all, the cost to bring a new antibiotic to market and keep it there is around $1.7 billion. Some experts are calling for a subscription model. Just as we pay a monthly fee to stream our favorite TV shows or read magazines, antibiotics, too, could be supported in a similar manner by countries around the world. The concept is known as "push-pull." Funding for basic research, early discovery, and preclinical studies on new drugs could be used to "push" new innovations into the antibiotic-discovery pipeline. Not-for-profit organizations like the Global Antibiotic Research and Development Partnership (GARDP) and Combating Antibiotic-Resistant Bacteria Biopharmaceutical Accelerator (CARB-X) are making important contributions to accelerating antibiotic research, but more is needed to secure antibiotics for the future.

The gap that comes after those promising early studies is known as the "valley of death"; this is where drugs go to die. The high expenses of early-phase clinical-drug development put this stage of research out of the grasp of academic researchers, and it doesn't make financial sense for drug companies to take on the massive cost burden to bring those promising compounds through the rigors of human testing, developing a commercial production chain, and gaining regulatory approval. Even after the drug is approved, there are extensive post-market monitoring costs ($350 million in the first ten years after approval). This second phase of testing in patients, production, and monitoring is where "pull" incentives could make a difference, encouraging companies to shepherd antibiotics through this costly process. The subscription model would ensure that research and development can recoup costs on a model that is decoupled from sales volume. This change would flip the pharmaceutical model on its head but also ensure a greater security against the looming crisis for us all.

As COVID-19 has shown us, we can't afford to wait; the global costs in terms of lives lost and economic crisis are too high. I wish it weren't the case, but the next COVID-19 might be around the corner. We don't know. And if it is, it might be an antibiotic-resistant superbug we have no weapons against. And next time, it could be even worse. We desperately need more resources to be prepared for what lies ahead. I continue to believe that plants offer the greatest chance for our survival, assisting existing medicines and even leading to the creation of new ones. That's why I go to work every day.

Epilogue

I sat at a small table in the center of our home garden. My children and I sorted through our late-summer harvest, the muggy heat of the day tapering off as evening settled all around us, with its chorus of buzzing insects and birdsong. Greenery abounded, thick vines bearing tomatoes, cucumbers, and sweet peas climbing upward along the tall bamboo poles we'd anchored to the ground, relying on the twine we'd tied them up with to carry the burden of their fruit-laden weight.

The garden perimeter was composed of a fanciful white picket fence; the entry gate was framed by a small arbor and decorated with the bold, colorful purple blooms of climbing clematis. Along the fence line, an array of bright wildflowers blossomed, attracting various pollinators—bees, hummingbirds—to visit. At the garden's edge, blackberry bushes reached as high as my chest, their woody stems decorated with vibrant green foliage and plump red-and-black fruit. Inside, wood chips from a fallen tulip poplar covered the paths that ran between the nineteen raised beds that Marco had built with the kids—each bed bursting with an array of edible, medicinal, and even poisonous plants.

There were tomatoes, okra, peppers, corn, sweet peas, yellow squash, leeks, chives, onions, cucumbers, butternut squash, zucchini, radishes, and assorted varieties of kale, lettuce, and cabbages we grew to supplement our groceries with fresh produce. Intermixed among the beds were the fragrant herbs—those wonderfully aromatic members of the mint family (mint, basil, rosemary, lavender, thyme, lemon balm, oregano) and the carrot family (parsley, cilantro). The medicines—also poisons, if misused—flourished in the special raised beds they'd been restricted to, so as to avoid confusion with the edible bounty of the garden. There was the greater celandine, motherwort, and foxglove—all of which the children were warned to never eat. Safer plant medicines (also used as medicinal foods) like marigold, purple coneflower (echinacea), borage, tulsi (also known as holy basil), catnip, and lemongrass burst forth from their confining clay pots. The biodiverse landscape we'd created in this small garden plot was a haven from the stress of the isolated and computerized world that the pandemic had created.

Each species held a special meaning to me—tied to memories of people I'd met, places I'd traveled to, where I'd experienced the food and medicinal properties of each firsthand. I noted some electric-blue flowers of borage that peeked out from the garden bed, the plant's greenery covered in fine hairs. It made me think of Zia Giovannina in Italy and the healing soup she made with chicken stock and the young stems and leaves of this herb to give to new mothers to help their milk come in. In the next garden bed, Bella clipped the fresh leafy stems of catnip with a pair of small scissors, tying small bundles at the base with twine as I'd taught her. These would be hung to dry for use throughout the year. I'd learned from Gorani mothers in Albania about the utility of catnip—which they call *strašnica*—to chase away nightmares and help children sleep by washing them in fragrant herbal baths and giving them honey-sweetened cups of its tea to drink.

The most recent addition to our family, my nephew, Trevor, made his way from the house out to the garden with a wicker basket in hand for us to load the day's harvest. Taking in the scene, I sighed with contentment.

On the opposite side of the garden, Giacomo was ripping at the plants again—a habit he'd picked up from watching me. He tore, crushed, and sniffed at leaves as he meandered through the rows of herbs and vegetables.

"Come here, MoMo," I called to him. He clutched a wad of leaves in his seven-year-old hand.

"Can I smell it, too?" I asked. He lifted his hands to my face, and I inhaled. He then held the bundle to his nose and did the same, mimicking my actions.

"It smells like lemons," he sighed.

I nodded. "But what does the plant look like?" I asked.

He brushed his fingers up and down the long, slender leaves emerging from the large clay pot as he considered the question. "It looks like grass."

"That's right. It's called lemongrass."

"Where did you learn about this plant, Mamma?" he asked.

"My friend, who was a very wise teacher and healer, taught me about this plant long ago. He lived in the rain forest."

I could see Don Antonio right then, his mischievous grin, holding a bundle of lemongrass, placing it on my chest for me to inhale its aroma as I lay in a palm-fiber hammock in the jungle fighting back the nausea of a stomach full of ant venom.

"I want to go to the rain forest," Giacomo said.

"Me, too, Ma," Donato said. "I want to go." At fifteen, his voice had deepened during his teenage growth spurt, and hearing this new version still surprised me. How had he, Trevor, and Bella, who would turn thirteen in just a couple weeks, grown taller than me already?

"We'll all go someday," I promised. Memories of the Amazon are ever present, an island floating somewhere in my mind, its calm black waters, flooded forests, and pink river dolphins calling to me in my dreams. There was so much more I needed to explore and learn from the plants and people who called that wonderland their home.

It was getting late. A firefly darted, and Trevor watched it. When the bug flashed again, he caught it in the cup of his hands. We gathered around to observe its tiny light flashing on and off through the gaps between his long fingers.

Marco came through the garden gate and joined us in our reverie. As we huddled together, in a moment of clarity, I realized that I was surrounded by everything I loved most in the world—all my children, my husband, all my plants. We'd built this little piece of paradise together.

Marco looked at me, and our eyes met. Mine misty with memory, his crinkled in a smile. How far we had come together since that eve when he played Romeo, climbing up to the balcony of my apartment in Italy. As if reading my thoughts, he leaned over and kissed me.

We started toward the house before he looked back to tell the kids, "Come on. Dinner is ready—I made lasagna!"

"Yum!" Giacomo shouted with glee. Donato grabbed the basket of garden veggies while Bella called out, "I'll make the salad!"

I took one last glance at the yard as we entered the house, glad to know I'll always have a garden right outside my door.

I'M A CHILD OF THE 1970s, a casualty of poisons used during the Vietnam War. While Agent Orange was intended to strip the forest trees bare of their dense foliage to improve visibility in the jungle as men played at war, it also stripped their children's future health away. Like me, thousands of children of US servicemen and Vietnamese villagers exposed to the potent defoliant Agent Orange were born with constellations of birth defects. Now in our forties, we share familiar health histories of missing bones, shortened limbs, and neural tube defects. While the US Veterans Health Administration included children of this era born with spina bifida in their coverage of health benefits for the children of Vietnam veterans in 1997, those like me who suffered from other life-altering bone defects were not covered. There was no assistance, no government health plan, to cover the hundreds of thousands of dollars in surgeries and prosthetic devices I've needed over my lifetime.

Nature's destruction comes at a cost. If there's anything to be learned from this pandemic or from the story of my life, it is that planetary health and human health are inextricably linked. Just as the oceans respond to the pull of the moon, nature finds a way to restore the balance of life as it is pulled to and fro. Ecosystem biodiversity is the key to the health of all life-forms. This is true whether one considers the myriad creatures that inhabit vast landmasses or the even microbes that live on and inside our bodies.

Just as bulldozers and chainsaws push their way deeper into rich forests, disrupting wildlife and creating new opportunities for disease to flourish, antibiotics, too, can disturb the balance for the humans and animals exposed to them. The dual pandemics of antibiotic-resistant infections and COVID-19 have made it plain that no one is safe if we continue on our current course. As I watch the death toll rise from COVID-19, I never forget that antibiotic-resistant infections continue to quietly claim the lives of more than seven hundred thousand people each year. In its first year, COVID-19 claimed the lives of two million people across the globe; antibiotic-resistant infections are projected to claim five times that—ten million lives per year—by 2050. How can we stand idly by as this disaster looms? No one is without risk, regardless of age, race, economic status, gender, or physical fitness. We need to prepare.

As human settlements expand with rising populations and governmental regulations are stripped away, wildlands are converted into agricultural lands, forests are felled for timber, and the ground is mined for precious metals and fossil fuels. All the while, biodiversity plummets across the United States and to the Amazon and beyond. As a result, at the very time that humankind is faced with some of the most daunting disease threats since the era of the Black Death, we are destroying the very resources that could be our salvation.

It's a challenge, to be sure, but we are not without hope. Nature holds the secrets to healing the planet and ourselves. Across the chasm of time, language, and culture, humans have persistently observed the interspecies behaviors of other animals and have experimented with nature's bountiful offerings. Although often relegated to the ranks of old wives' tales by Western scientists, this rich bounty of knowledge concerning the uses of botanical ingredients as medicine is anything but. These traditional systems of medicine represent a natural pharmacy that has matured over millennia, refined and improved upon with each generation. While modern scientists don't yet fully understand which plant medicines are the most effective, or how they work or which compounds are most responsible, we do know that some of our most essential pharmaceutical medicines used across the globe were initially discovered in such plants.

Like our earliest ancestors leaving the jungle, like the first men and women to build ships and set sail on the roiling ocean, we must continue our exploration. Nature is all around us; it is within us. We are nature. We must remember our place in the natural world and work with it to discover new ways for it to help us as we help it. Our fates are intertwined just like the strands of our DNA.

Nature never fails to amaze me. Its complexity, its beauty, its genius. How it finds a way to thrive and adapt, and what secrets it contains. I need look no further than my family's little garden out back to be reminded of nature's rich offerings and our duty to be good stewards to these plants that have so much to offer us. Nature is my home. And I know one thing for sure: I will not cease my exploration.

Acknowledgments

To my husband, Marco, you are my rock and partner in life. Your dedication to our family and your encouragement and support of my dreams made their achievement possible. I couldn't have made this journey or written this book without you. I can't wait to experience the adventures that await us in the future—hopefully somewhere over a crystalline sea by sailboat.

I would not be here to tell this story without the love and guidance of my parents. Momma, you instilled a fortitude that allowed me to conquer what others would see as impossible. You never let me feel disabled, and always showed me that the only limitations to my success were my will and effort to push forward to succeed. Thank you for all of the months of your life spent by my side at the doctor's office and the hospital bed. I couldn't have done it without you. Daddy, your creativity and work ethic inspired me to think, dream, and do bigger things—to stretch my imagination past points I'd never thought possible. You taught me to laugh and love with an open heart, and to follow the path I was meant for, even if it was the one less traveled. Thank you for all of the times we spent in the woods together, giving me the space and time to fall in love with the natural world.

To my children—Donato, Bella, and Giacomo, and my nephew, Trevor—you are my heart and joy. Each one of you has brought something beautiful to this world, and I'm so thankful to be a part of your lives.

To my sister, who enriched my childhood with outdoor adventures climbing trees and running barefoot in the muck under the Florida sunshine. To Granny, who, until she passed at the age of ninety-eight, continued to share her spirit of adventure and travel with me. Who knows, maybe I'll start riding a motorcycle cross-country after retirement someday just like she did!

To my wonderful family in Italy—thank you to my parents-in-law, Milagros and Donato, for supporting my research through the earliest days of fieldwork, and being patient with me even when my plant dryers overheated the cantina and ruined the barrels of wine. To my brothers- and sisters-in-law, and my wonderful nieces and nephews—thank you for all the ways that you have enriched my life and for all of the times that you joined me in the Italian countryside to help me harvest wild plants.

To Dr. Chad Price and all the doctors who healed and rebuilt my growing body throughout childhood, as well as my leg men (prosthetists Charlie and Will)—thanks for making my adventures by land and sea physically possible with the tools I have today!

I owe so much to my mentors, and I want to especially recognize Drs. Michelle Lampl, Larry Wilson, Brad Bennett, Bob Swerlick, Lisa Plano, Mark Smeltzer, and Cesar Compadre, who not only taught me what it is to be a scientist but also helped form the mentorship style I use today as a professor. Thank you to Dr. Ray Schinazi and Pete McTier for your counsel, support, and encouragement as my career has developed. You have all inspired me, pushed me to reach a higher potential, and advocated for me when I needed it most, and I am incredibly appreciative of your time and counsel.

I am grateful to my earliest mentors from the days of training in the hospital microbiology labs and the emergency room, and the many paramedics, lab technicians, nurses, and doctors who taught me how to stand strong in the face of crises and to merge empathy and compassion with the practice of medicine. I especially want to thank Ms. Betty, Tracy Camp, Stephanie Chisholm-Beasley, and Drs. James Meade, Steven Mishkind, and William Crankshaw. Thank you to Bill Stanko for bringing the science fair to life in our small corner of Florida, and for coaching me as I went on to the state and international competitions.

I am deeply grateful to the many communities that have welcomed me into their homes and shared their knowledge of nature and medicine with me. My life has been so incredibly enriched through these experiences in learning the meaning of health through different cultural lenses. I want to especially thank Don Antonio, Zia Elena, Zia Giovannina, and all of the many healers who have taken the time to share their wisdom with me; I am eternally grateful.

To my scientific collaborators across the globe, you have not only been integral to the research described in this book and published in the scientific literature but have also been amazing peer mentors. Your scientific talent, work ethic, curiosity, and insight have both inspired and guided me on this journey. The colleagues I've had the great fortune to work with are too numerous to list, but in particular I want to recognize Drs. Andrea Pieroni, Alex Horswill, Avni Hajdari, Alessandro Saitta, Kier Klepzig, Alfonso La Rosa, Christian Melander, Dan Zurawski, Julia Kubanek, Rheinallt Jones, and Tim Read for your continued collaboration over the years.

Thank you to my collaborators in science communication who've helped me spread my love for plants and pharmacology through new mediums in print, audio, and video. Thanks to the talented journalists, photographers, filmmakers, and others who have journeyed with me into the field and lab to report on our research and share my research team's curiosity and joy for exploration and discovery with the rest of the world. I want to especially recognize Carol Clark, Ferris Jabr, Maryn McKenna, Rob Cohen, and Christine Roth.

To my research team, I have been so fortunate to work with you in both the field and the lab. Your energy and curiosity are the lifeblood of the lab, and there has been no greater joy for me as a teacher, mentor, and boss than to see how many of you have continued in your own scientific journeys to great heights. It is an impossible task to fully recognize all of your contributions to this research, but I want to especially acknowledge Drs. Thara Samarakoon, James Lyles, Gina Porras, François Chassagne, Sunmin Woo, Kerac Falk, Emily Gurnee, and Akram Salam, as well as Kate Nelson, Brandon Dale, Micah Dettweiler, Monique Salazar, Patty Calarese, Myleshia Perryman, Marco Caputo, Lewis Marquez, Caitlin Risener, Bledar Pulaj, Faraz Khan, Huaqiao Tang, Fabien Schultz, Amelia Muhs, Nick Richwagen, Xena Huang, Mickie Xu, Danielle Carrol, Rozenn Pineau, Sarah Hanson, and the more than one hundred undergraduate interns and visiting scholars who I've had the pleasure of working with. No matter the years that pass, you will always be welcomed home in the Quave Lab.

Over the span of my training and career, there have been a number of peer mentors in the fields of ethnobotany and pharmacognosy—beautiful friends and scientific colleagues—who have banded together to lift one another up, to listen, to counsel, to encourage, and to commiserate. To Drs. Janna Rose, Janelle Baker, Nanci Ross, Sunshine Brosi, Ina Vandebroek, Rick Stepp, Nadja Cech, Sandra Loesgen, John de la Parra, and Sonia Peter, you continue to inspire me every day, and I eagerly await our annual gatherings. I am indebted to my dear friends and colleagues Drs. Maria Fadiman and Susanne Masters, who so generously read many revisions of this book and cheered me on, lifting me up even on the days when I thought all was lost. And to Dr. Justin Nolan, your early departure from this world has left a gaping hole in our hearts. I count myself incredibly fortunate to have known you as my treasured friend and kindred spirit.

Science is simply not possible without the funding support necessary to drive the work. I am grateful to all of the funding bodies that have supported me during my career, and especially those that took a chance on me early on. I want to especially recognize the Emory College International Scholars Program, the Foundation for Science and Disability, the National Kappa Alpha Theta Foundation, the National Institutes of Health (NIH) National Center for Complementary and Alternative Medicine/National Center for Complementary and Integrative Health, the NIH National Institute of Allergy and Infectious Disease, the US Department of Agriculture, Botany in Action, the Garden Club of America, the Winship Melanoma Fund, the Samueli Foundation, the Community Foundation for Greater Atlanta, the Estate of Hubert Henry Whitlow, Jr., the Jones Center at Ichauway, and The Marcus Foundation. Beyond large funding bodies, my work has been helped tremendously by the patronage of individuals passionate about this work. Whether you gave $10 or $10,000, I am forever grateful and honored by your contributions to the lab and the herbarium. Our research and outreach initiatives wouldn't be possible without you.

To my lifelong friends, we have been at one another's sides through childhood, college, marriages, births, illness, deaths—the adventures, joys, and tragedies that come with the ebb and flow of life. There are no others I'd rather take this journey with than you. To Biggie, Brian, Jayme, Julius, Jenny, Dan, Robyn, the EOME crew (Sahil, Katie, Kenny, Cindy, Matt, Jen, August, Liz, Hunter, and Vanessa), and the Arkansas girls, thank you for the laughter and joy that you bring to my life. Thank you especially to my treasured friends Mandy Fugate and Jen Waters, who gave so generously of their time reading and providing feedback on the book!

Thank you to my fantastic literary agent, Elias Altman, for first believing that this is a story that should be told. Thank you to my fabulous editor, Georgia Bodnar, Gretchen Schmid, and the incredibly talented team at Viking for your insight and advice on how to craft my writing into a tale that will endure the years.

My life has been touched by so many amazing people, and I apologize if I wasn't able to name you here.

Appendices

Schinus terebinthifolia

Other Resources

How you can make a difference:

If you would like to support the research on anti-infective drug discovery from nature discussed in this book, please consider making a donation to my research group. Funds are used to support the salaries of student trainees and research staff, laboratory supplies, and research expedition expenses. Donations are tax deductible in the United States and can be easily made via a secure portal at my university. More information can be found on my research website (https://etnobotanica.us/donate) on how to make a contribution.

Additional learning resources:

Check out other video and audio content related to the topics discussed in the book through the following portals:

- Teach Ethnobotany YouTube channel (www.youtube.com/user/Teach Ethnobotany). I created this channel to disseminate videos of lectures and short presentations on the topics of ethnobotany and health. This includes videos of my own lectures and also recordings of presentations from scientific meetings of the Society for Economic Botany and Society of Ethnobiology.
- *Foodie Pharmacology* podcast (https://foodiepharmacology.com/). I am the co-creator and host of a podcast dedicated to exploring the science

behind food and health. Episodes include interviews with experts on the topics of food, health, culture, pharmacology, farming, and more.

- Integrated Digitized Biocollections (iDigBio; www.idigbio.org). A training program dedicated to the digitization (photography and data entry) of natural history collections, including herbaria, supported by the National Science Foundation.
- Convention on International Trade in Endangered Species of Wild Fauna and Flora (https://cites.org/eng). An international agreement between governments with the aim of ensuring that the international trade in wild animal and plant species doesn't threaten species survival.
- Global Antibiotic Research and Development Partnership (https://gardp.org); Combating Antibiotic-Resistant Bacteria Biopharmaceutical Accelerator (CARB-X; https://carb-x.org/). Not-for-profit organizations that are making important contributions to accelerating antibiotic research.
- Southeast Regional Network of Expertise and Collections (SERNEC; https://sernecportal.org/portal/). A consortium of herbaria from the southeastern United States that provides herbarium specimen images and metadata for research, management planning, and informing the public.

Resources for students and scientists:

One of the best things that happened early in my career was meeting others who shared my love of the connections between humankind and nature. We are fortunate to have several international scientific societies dedicated to the ethical study of the human-nature interface as well as the study of the medicinal potential of plants. These are great organizations to become connected with if you are interested in pursuing studies in this field. Moreover, they are all incredibly welcoming of students and newcomers, and I encourage you to consider engaging with members online and by attending one of their society conferences. Here are some that I recommend:

- The Society for Economic Botany (www.econbot.org) is about people exploring the uses of, and our relationship with, plants, cultures, and our environment—plants and human affairs.
- The American Society of Pharmacognosy (www.pharmacognosy.us) is dedicated to discovering nature's molecular potential. This includes research on plants but also many other natural sources, such as marine organisms, soil microbes, fungi, and more.
- The Society of Ethnobiology (https://ethnobiology.org/) is an organization of scholars, activists, and communities promoting and supporting the interdisciplinary study of past and present relationships between people and the environment.
- The International Society of Ethnobiology (www.ethnobiology.net) a global, collaborative network of individuals and organizations working to preserve vital links between human societies and the natural world.

Index of Botanical and Fungal Names

Family names are denoted in capital letters. Scientific names (genus and species) are presented in italics and common names in normal font. Family assignments follow the Angiosperm Phylogeny Group IV and MycoBank assignments. Throughout the book, the author epithet of each species has been left off to improve the reading flow of the text. Full accepted names of all of the species included in the book can be accessed at www.worldfloraonline.org for plants and www.mycobank.org for fungi.

List of Acronyms and Initialisms

ACE2: Angiotensin-Converting Enzyme 2

AKA: Above-knee amputee

AMR: Antimicrobial resistance

APC: Armored personnel carrier

BKA: Below-knee amputee

BSL: Biosafety level

CA-MRSA: Community-associated methicillin-resistant *Staphylococcus aureus*

CDC: Centers for Disease Control and Prevention

CENaP: Center for Ethnobiology and Natural Products

CF: Cystic fibrosis

COVID-19: Coronavirus disease 2019

CRAB: Carbapenem-resistant *Acinetobacter baumannii*

CSHH: Center for the Study of Human Health

DMSO: Dimethyl sulfoxide

DMT: Dimethyltryptamine

ESKAPE: *Enterococcus faecium, Staphylococcus aureus, Klebsiella pneumoniae, Acinetobacter baumanni, Pseudomonas aeruginosa,* and *Enterobacter* species

FDA: Food and Drug Administration

FIU: Florida International University

GPS: Global positioning system

HA-MRSA: Healthcare-associated methicillin-resistant *Staphylococcus aureus*

MAOI: Monoamine oxidase inhibitor

MERS: Middle East respiratory syndrome coronavirus

MRSA: Methicillin-resistant *Staphylococcus aureus*

NCCIH: National Center for Complementary and Integrative Health

NGO: Nongovernmental organization

NIAID: National Institute of Allergy and Infectious Diseases

NIH: National Institutes of Health

OSN: Open Science Network

PI: Principal investigator

PPE: Personal protective equipment

QNPL: Quave Natural Products Library

R & D: Research and development

SEB: Society for Economic Botany

SERNEC: Southeast Regional Network of Expertise and Collections

SSEF: State Science and Engineering Fair

SSRI: Selective serotonin reuptake inhibitor

TB: Tuberculosis

UALR: University of Arkansas at Little Rock

UAMS: University of Arkansas for Medical Sciences

USDA: United States Department of Agriculture

WHO: World Health Organization

Notes

Prologue

4 **one species native to Italy:** Cassandra L. Quave, Lisa R. W. Plano, Traci Pantuso, and Bradley C. Bennett, "Effects of Extracts from Italian Medicinal Plants on Planktonic Growth, Biofilm Formation and Adherence of Methicillin-Resistant *Staphylococcus aureus*," *Journal of Ethnopharmacology* 118, no. 3 (August 2008): 418–28, https://doi.org/10.1016/j.jep.2008.05.005; Cassandra L. Quave, Miriam Estévez-Carmona, Cesar M. Compadre, Gerren Hobby, Howard Hendrickson, Karen E. Beenken, and Mark S. Smeltzer, "Ellagic Acid Derivatives from *Rubus ulmifolius* Inhibit *Staphylococcus aureus* Biofilm Formation and Improve Response to Antibiotics," *PLoS One* 7 (January 2012): e28737, https://doi.org/10.1371/journal.pone.0028737; Benjamin M. Fontaine, Kate Nelson, James T. Lyles, Parth B. Jariwala, Jennifer M. García-Rodriguez, Cassandra L. Quave, and Emily E. Weinert, "Identification of Ellagic Acid Rhamnoside as a Bioactive Component of a Complex Botanical Extract with Anti-biofilm Activity," *Frontiers in Microbiology* 8 (March 2017): 496, https://doi.org/10.3389/fmicb.2017.00496.

4 **More than seven hundred thousand people:** J. O'Neill, *Review on Antimicrobial Resistance*, 2016, Wellcome Trust and the UK Department of Health, London.

5 **Even scarier is that since the 1980s:** Lynn L. Silver, "Challenges of Antibacterial Discovery," *Clinical Microbiology Reviews* 24, no. 1 (January 2011): 71–109, https://doi.org/10.1128/cmr.00030-10.

5 **out of the fifteen new antibiotics:** Maryn McKenna, "The Antibiotics Business Is Broken—but There's a Fix," *Wired*, April 25, 2019, www.wired.com/story/the-antibiotics-business-is-broken-but-theres-a-fix/.

6 **In 1900 in the United States:** Centers for Disease Control and Prevention, "Achievements in Public Health, 1900–1999: Control of Infectious Diseases," *Morbidity and Mortality Weekly Report* 48, no. 29 (July 1999): 621–29.

7 **In a 1945 speech:** "Penicillin's Finder Assays Its Future," *New York Times*, June 26, 1945.

8 **Of the estimated 374,000 species:** Maarten J. M. Christenhusz and James W. Byng, "The Number of Known Plants Species in the World and Its Annual Increase," *Phytotaxa* 261, no. 3 (May 2016): 201–17, https://doi.org/10.11646/phytotaxa.261.3.1.

8 **medicinal use of at least 33,443:** Medicinal Plant Names Services, Kew Science, accessed March 29, 2021, https://mpns.science.kew.org/mpns-portal/.

9 **A Nobel Prize was even awarded:** Xin-Zhuan Su and Louis H. Miller, "The Discovery of Artemisinin and the Nobel Prize in Physiology or Medicine," *Science China Life Sciences* 58 (2015): 1175–79, https://doi.org/10.1007/s11427-015-4948-7.

10 **The term *ethnobotany* was first coined:** J. W. Harshberger, "Purposes of Ethnobotany," *Botanical Gazette* 21, no. 3 (1896): 146–54.

10 **ethnobotany is the science:** Ghillean T. Prance, "Ethnobotany, the Science of Survival: A Declaration from Kaua'i," *Economic Botany* 61 (2007): 1–2, https://doi.org/10.1007/BF02862367.

11 **Staph infections kill more than:** Athena P. Kourtis, Kelly Hatfield, James Baggs, Yi Mu, Isaac See, Erin Epson, Joelle Nadle, Marion A. Kainer, Ghinwa Dumyati, Susan Petit, Susan M. Ray, Emerging Infections Program MRSA author group, David Ham, Catherine Capers, Heather Ewing, Nicole Coffin, L. Clifford McDonald, John Jernigan, and Denise Cardo, "Vital Signs: Epidemiology and Recent Trends in Methicillin-Resistant and in Methicillin-Susceptible *Staphylococcus aureus* Bloodstream Infections—United States," *Morbidity and Mortality Weekly Report* 68, no. 9 (March 2019): 214–19, https://doi.org/10.15585/mmwr.mm6809e1.

Chapter 1: My Leg and the Wilderness

25 **strain of *E. coli* had been discovered:** Ji Youn Lim, Jangwon Yoon, and Carolyn J. Hovde, "A Brief Overview of *Escherichia coli* O157:H7 and Its Plasmid O157," *Journal of Microbiology and Biotechnology* 20, no. 1 (January 2010): 5–14.

Chapter 2: Welcome to the Jungle

40 **The birch polypore:** Luigi Capasso, "5300 Years Ago, the Ice Man Used Natural Laxatives and Antibiotics," *Lancet* 352, no. 9143 (December 1998): 1864, https://doi.org/10.1016/s0140-6736(05)79939-6.

40 **seven hundred plant formulas and remedies:** Felix Cunha, "I. The Ebers Papyrus," *American Journal of Surgery* 77 no. 1 (January 1949): 134–36, https://doi.org/10.1016/0002-9610(49)90394-3; Anke Hartmann, "Back to the Roots—Dermatology in Ancient Egyptian Medicine," *Journal der Deutschen Dermatologischen Gesellschaft* 14, no. 4 389–96, https://doi.org/https://doi.org/10.1111/ddg.12947.

48 **In 2012, the resin of this dragon's blood:** Q. M. Yeo, Rustin Crutchley, Jessica Cottreau, Anne Tucker, and Kevin W. Garey, "Crofelemer, a Novel Antisecretory Agent Approved for the Treatment of HIV-Associated Diarrhea," *Drugs Today (Barc)* 49, no. 4 (April 2013): 239–52, https://doi.org/10.1358/dot.2013.49.4.1947253; Poorvi Chordia and Roger D. MacArthur, "Crofelemer, a Novel Agent for Treatment of Non-infectious Diarrhea in HIV-Infected Persons," *Expert Review of Gastroenterology and Hepatology* 7, no. 7 (2013): 591–600, https://doi.org/10.1586/17474124.2013.832493.

53 **When ayahuasca is orally ingested with MAOIs:** Elisabet Domínguez-Clavé, Joaquim Soler, Matilde Elices, Juan C. Pascual, Enrique Álvarez, Mario de la Fuente Revenga, Pablo Friedlander, Amanda Feilding, and Jordi Riba, "Ayahuasca: Pharmacology, Neuroscience and Therapeutic Potential," *Brain Research Bulletin* 126 (September 2016): 89–101, https://doi.org/https://doi.org/10.1016/j.brainresbull.2016.03.002.

Chapter 3: Worms in the Belly

77 **"Sickness is not just an isolated event":** Nancy Scheper-Hughes and Margaret M. Lock, "The Mindful Body: A Prolegomenon to Future Work in Medical Anthropology," *Medical Anthropology Quarterly* 1, no. 1 (March 1987): 6–41, https://doi. org/https://doi.org/10.1525/maq.1987.1.1.02a00020.

Chapter 4: An Unexpected Houseguest

84 **It wasn't until 2010:** "The Nagoya Protocol on Access and Benefit-Sharing," Convention on Biological Diversity, updated March 18 2021, www.cbd.int/abs/.

93 **Communities were losing:** Andrea Pieroni, Cassandra Quave, Sabine Nebel, and Michael Heinrich, "Ethnopharmacy of the Ethnic Albanians (Arbëreshë) of Northern Basilicata, Italy," *Fitoterapia* 73, no. 3 (June 2002): 217–41, https://doi.org/ 10.1016/s0367-326x(02)00063-1.

93 **For the Arbëreshë, wild greens:** Andrea Pieroni, Sabine Nebel, Cassandra Quave, Harald Münz, and Michael Heinrich, "Ethnopharmacology of Liakra: Traditional Weedy Vegetables of the Arbëreshë of the Vulture Area in Southern Italy," *Journal of Ethnopharmacology* 81, no. 2 (2002): 165–85, https://doi.org/10.1016/ s0378-8741(02)00052-1.

95 **science of diet, lifestyle, health:** Seven Countries Study, accessed April 3, 2021, www.sevencountriesstudy.com.

100 **There were many spiritual ailments:** Cassandra L. Quave and Andrea Pieroni, "Ritual Healing in Arbëreshë Albanian and Italian Communities of Lucania, Southern Italy," *Journal of Folklore Research* 42, no. 1 (2005): 57–97.

Chapter 5: Wash and Fold

134 **Fairchild was responsible:** Daniel Stone, *The Food Explorer: The True Adventures of the Globe-Trotting Botanist Who Transformed What America Eats* (New York: Dutton, 2018).

135 **108,345 cases of invasive MRSA:** Centers for Disease Control and Prevention, *Active Bacterial Core Surveillance Report, Emerging Infections Program Network, Methicillin-Resistant Staphylococcus aureus 2006*, last updated January 30, 2012, https://www.cdc. gov/abcs/reports-findings/survreports/mrsa06.pdf.

135 **14,627 deaths from AIDS:** Centers for Disease Control and Prevention, *HIV/ AIDS Surveillance Report: Cases of HIV Infection and AIDS in the United States and Dependent Areas, 2006*, 2008, www.cdc.gov/hiv/pdf/statistics_2006_HIV_Surveillance_Re port_vol_18.pdf.

135 **But even athletes:** Eugene Brent Kirkland and Brian B. Adams, "Methicillin-Resistant *Staphylococcus aureus* and Athletes," *Journal of the American Academy of Dermatology* 59, no. 3 (September 1, 2008): 494–502, https://doi.org/10.1016/j.jaad. 2008.04.016.

135 **infections tied to the gym:** Philip R. Cohen, "The Skin in the Gym: A Comprehensive Review of the Cutaneous Manifestations of Community-Acquired Methicillin-Resistant *Staphylococcus aureus* Infection in Athletes," *Clinical Dermatology* 26, no. 1 (January 2008): 16–26, https://doi.org/10.1016/j.clindermatol.2007.10.006.

135 **In 2004, a healthy eighteen-year-old:** Ian Sample, "Bacteria Tests Reveal How MRSA Strain Can Kill in 24 Hours," *Guardian*, January 19, 2007, www.theguardian.com/society/2007/jan/19/health.medicineandhealth2.

136 **In 2007, a seventeen-year-old:** Mythili Rao and Tim Langmaid, "Bacteria That Killed Virginia Teen Found in Other Schools," CNN, October 18, 2007, www.cnn.com/2007/HEALTH/10/18/mrsa.cases/.

136 **could lead to necrotizing pneumonia:** Yves Gillet, Philippe Vanhems, Gerard Lina, Michèle Bes, François Vandenesch, Daniel Floret, and Jerome Etienne, "Factors Predicting Mortality in Necrotizing Community-Acquired Pneumonia Caused by *Staphylococcus aureus* Containing Panton-Valentine Leukocidin," *Clinical Infectious Diseases* 45, no. 3 (August 2007): 315–21, https://doi.org/10.1086/519263.

137 **Numerous clinical trials have shown:** Erin I. Armentrout, George Y. Liu, and Gisláine A. Martins, "T Cell Immunity and the Quest for Protective Vaccines against *Staphylococcus aureus* Infection," *Microorganisms* 8, no. 12 (December 2020): 1936, https://doi.org/10.3390/microorganisms8121936.

Chapter 6: From the Field to the Lab

156 **116 different topical remedies:** Cassandra L. Quave, Andrea Pieroni, and Bradley C. Bennett, "Dermatological Remedies in the Traditional Pharmacopoeia of Vulture-Alto Bradano, Inland Southern Italy," *Journal of Ethnobiology and Ethnomedicine* 4, article 5 (2008), https://doi.org/10.1186/1746-4269-4-5.

Chapter 7: Babies and Biofilms

188 **Abscesses are surprisingly common:** Breena R. Taira, Adam J. Singer, Henry C. Thode Jr., and Christopher C. Lee, "National Epidemiology of Cutaneous Abscesses: 1996 to 2005," *American Journal of Emergency Medicine* 27, no. 3 (March 2009): 289–92, https://doi.org/10.1016/j.ajem.2008.02.027.

197 **It can take ten to twenty years:** Olivier J. Wouters, Martin McKee, and Jeroen Luyten, "Estimated Research and Development Investment Needed to Bring a New Medicine to Market, 2009–2018," *Journal of the American Medical Association* 323, no. 9 (March 2020): 844–53, https://doi.org/10.1001/jama.2020.1166.

197 **there is an estimated $350 million:** John H. Rex, "Melinta, Part 2/Bankruptcy Is Not the End/Post-approval Costs for an Antibiotic," Antimicrobial Resistance Solutions, January 20, 2020, https://amr.solutions/2020/01/07/melinta-part-2-bankruptcy-is-not-the-end-post-approval-costs-for-an-antibiotic/.

Chapter 8: A Lab of My Own

208 **Van Leeuwenhoek's famous 1676 paper:** Antoni Van Leeuwenhoek, "Observations, Communicated to the Publisher by Mr. Antony van Leewenhoeck, in a Dutch Letter of the 9th Octob. 1676. here English'd: Concerning Little Animals by him Observed in Rain-Well-Sea- and Snow Water; as also in Water wherein Pepper had Lain Infused," *The Royal Society Philosophical Transactions*, March 25, 1677, https://doi.org/10.1098/rstl.1677.0003.

Chapter 9. The Sea Cabbage

230 **To be a hot spot:** "Biodiversity Hotspots Defined," Critical Ecosystem Partnership Fund, accessed April 3, 2021, www.cepf.net/our-work/biodiversity-hotspots/hotspots-defined.

230 **An incredible 44 percent of all plants:** Norman Myers, Russell A. Mittermeier, Cristina G. Mittermeier, Gustavo A. B. da Fonseca, and Jennifer Kent, "Biodiversity Hotspots for Conservation Priorities," *Nature* 403 (February 2000): 853–58, https://doi.org/10.1038/35002501.

230 **We're in a race to collect:** Melanie-Jayne R. Howes, Cassandra L. Quave, Jérôme Collemare, Evangelos C. Tatsis, Danielle Twilley, Ermias Lulekal, Andrew Farlow, Liping Li, María-Elena Cazar, Danna J. Leaman, Thomas A. K. Prescott, William Milliken, Cathie Martin, Marco Nuno De Canha, Namrita Lall, Haining Qin, Barnaby E. Walker, Carlos Vásquez-Londoño, Bob Allkin, Malin Rivers, Monique S. J. Simmonds, Elizabeth Bell, Alex Battison, Juri Felix, Felix Forest, Christine Leon, China Williams, and Eimear Nic Lughadha, "Molecules from Nature: Reconciling Biodiversity Conservation and Global Healthcare Imperatives for Sustainable Use of Medicinal Plants and Fungi," *Plants, People, Planet* 2, no. 5 (September 2020): 463–81, https://doi.org/https://doi.org/10.1002/ppp3.10138.

231 **Medicinal plants constitute the primary pharmacopoeia:** Molly Meri Robinson and Xiaorui Zhang, *The World Medicines Situation 2011—Traditional Medicines: Global Situation, Issues and Challenges* (World Health Organization, Geneva: 2011).

232 **Of the three main Egadi Islands:** William Henry Smyth, *Memoir Descriptive of the Resources, Inhabitants and Hydrography of Sicily and Its Islands, Interspersed with Antiquarian and Other Notices* (London: John Murray, Albemarle-Street, 1824), 244.

232 **The island has a rich history:** Leonardo Orlandini, "Trapani succintamente descritto del Canonico Leonardo Orlandini," trans. Gino Lipari, Trapani Nostra, www.trapaninostra.it/libri/Gino_Lipari/Trapani_Succintamente_Descritto/Trapani_Succintamente_Descritto.pdf; Giovanni Andrea Massa, *La Sicilia in Prospettiva. Parte Seconda. Cioè Le Citta, Castella, Terre e Luoghi Esistenti e Non Esistenti in Sicilia, la Topografia Littorale, li Scogli, Isole e Penisole Intorno ad Essa. Esposti in Veduta Da Un Religioso Della Compagnia Di Gesù* (Palermo: Stamparia di Francesco Cichè, 1709).

233 **This and other members of the *Daphne* genus:** Timur Tongur, Naciye Erkan, and Erol Ayranci, "Investigation of the Composition and Antioxidant Activity of Acetone and Methanol Extracts of *Daphne sericea* L. and *Daphne gnidioides* L.," *Journal of Food Science and Technology* 55, no. 4 (April 2018): 1396–1406, https://doi.org/10.1007/s13197-018-3054-9.

234 **This lack of biodiversity:** S. C. Cunningham, R. Mac Nally, P. J. Baker, T. R. Cavagnaro, J. Beringer, J. R. Thomson, and R. M. Thompson, "Balancing the Environmental Benefits of Reforestation in Agricultural Regions," *Perspectives in Plant Ecology, Evolution and Systematics* 17, no. 4 (July 2015): 301–17, https://doi.org/10.1016/j.ppees.2015.06.001.

244 **I had become fascinated with the *Daphne*:** Cassandra L. Quave and Alessandro Saitta, "Forty-Five Years Later: The Shifting Dynamic of Traditional Ecological Knowledge on Pantelleria Island, Italy," *Economic Botany* 70 (December 2016): 380–93, https://doi.org/10.1007/s12231-016-9363-x.

250 **possible links to neurodegenerative diseases:** Fiona Q. Bui, Cassio Luiz Coutinho Almeida-da-Silva, Brandon Huynh, Alston Trinh, Jessica Liu, Jacob

Woodward, Homer Asadi, and David M. Ojcius, "Association between Periodontal Pathogens and Systemic Disease," *Biomedical Journal* 42, no. 1 (February 2019): 27–35, https://doi.org/https://doi.org/10.1016/j.bj.2018.12.001; Mahtab Sadrameli, Praveen Bathini, and Lavinia Alberi, "Linking Mechanisms of Periodontitis to Alzheimer's Disease," *Current Opinion in Neurology* 33, no 2 (April 2020): 230–38.

250 **Eleven proved efficacious:** Danielle H. Carrol, François Chassagne, Micah Dettweiler, and Cassandra L. Quave, "Antibacterial Activity of Plant Species Used for Oral Health against *Porphyromonas gingivalis*," *PLoS One* 15, no. 10 (October 2020): e0239316, https://doi.org/10.1371/journal.pone.0239316.

250 **Micah's undergraduate honors research:** Micah Dettweiler, James T. Lyles, Kate Nelson, Brandon Dale, Ryan M. Reddinger, Daniel V. Zurawski, and Cassandra L. Quave, "American Civil War Plant Medicines Inhibit Growth, Biofilm Formation, and Quorum Sensing by Multidrug-Resistant Bacteria," *Scientific Reports* 9 (May 2019): 7692, https://doi.org/10.1038/s41598-019-44242-y.

250 **There was a surge in antibiotic-resistant infections:** Centers for Disease Control and Prevention, "*Acinetobacter baumannii* Infections among Patients at Military Medical Facilities Treating Injured U.S. Service Members, 2002–2004," *Morbidity and Mortality Weekly Report* 53, no. 45 (November 2004): 1063–66; Callie Camp and Owatha L. Tatum, "A Review of *Acinetobacter baumannii* as a Highly Successful Pathogen in Times of War," *Laboratory Medicine* 41, no. 11 (November 2010): 649–57, https://doi.org/10.1309/LM90IJNDDDWRI3RE.

250 *Acinetobacter* **is exceptionally adept:** S. I. Getchell-White, L. G. Donowitz, and D. H. Gröschel, "The Inanimate Environment of an Intensive Care Unit as a Potential Source of Nosocomial Bacteria: Evidence for Long Survival of *Acinetobacter calcoaceticus*," *Infection Control and Hospital Epidemiology* 10, no. 9 (August 1989): 402–7, https://doi.org/10.1086/646061; A. Jawad, A. M. Snelling, J. Heritage, and P. M. Hawkey, "Exceptional Desiccation Tolerance of *Acinetobacter radioresistens*," *Journal of Hospital Infection* 39, no. 3 (July 1998): 235–40, https://doi.org/10.1016/s0195-6701(98)90263-8.

251 **contaminated surface (such as a bed rail):** M. Catalano, L. S. Quelle, P. E. Jeric, A. Di Martino, and S. M. Maimone, "Survival of *Acinetobacter baumannii* on Bed Rails during an Outbreak and during Sporadic Cases," *Journal of Hospital Infection* 42, no. 1 (May 1999): 27–35, https://doi.org/10.1053/jhin.1998.0535.

251 **CRAB hospitalized 8,500 people:** Centers for Disease Control and Prevention, *Antibiotic Resistance Threats in the United States, 2019,* December 2019, www.cdc.gov/drugresistance/pdf/threats-report/2019-ar-threats-report-508.pdf.

251 **pentagalloyl glucose, or PGG:** Micah Dettweiler, Lewis Marquez, Michelle Lin, Anne M. Sweeney-Jones, Bhuwan Khatri Chhetri, Daniel V. Zurawski, Julia Kubanek, and Cassandra L. Quave, "Pentagalloyl Glucose from *Schinus terebinthifolia* Inhibits Growth of Carbapenem-Resistant *Acinetobacter baumannii*," *Scientific Reports* 10 (September 2020): 15340, https://doi.org/10.1038/s41598-020-72331-w.

Chapter 10. Billy Fell off the Swing

255 **TB still kills 1.4 million:** "Tuberculosis," World Health Organization, October 14, 2020, www.who.int/news-room/fact-sheets/detail/tuberculosis.

257 **multidrug-resistant ESKAPE pathogens:** Helen W. Boucher, George H. Talbot, John S. Bradley, John E. Edwards, David Gilbert, Louis B. Rice, Michael

Scheld, Brad Spellberg, and John Bartlett, "Bad Bugs, No Drugs: No ESKAPE! An Update from the Infectious Diseases Society of America," *Clinical Infectious Diseases* 48, no. 1 (January 2009): 1–12, https://doi.org/10.1086/595011.

258 **dangerous to children with cystic fibrosis:** Cystic Fibrosis Foundation, accessed April 5, 2021, www.cff.org/What-is-CF/About-Cystic-Fibrosis.

261 **Collections of plant specimens:** L. Alan Prather, Orlando Alvarez-Fuentes, Mark H. Mayfield, and Carolyn J. Ferguson, "The Decline of Plant Collecting in the United States: A Threat to the Infrastructure of Biodiversity Studies," *Systematic Botany* 29, no. 1 (January 2004): 15–28; Daniel P. Bebber, Mark A. Carine, John R. I. Wood, Alexandra H. Wortley, David J. Harris, Ghillean T. Prance, Gerrit Davidse, Jay Paige, Terry D. Pennington, Norman K. B. Robson, and Robert W. Scotland, "Herbaria Are a Major Frontier for Species Discovery," *Proceedings of the National Academy of Sciences* 107, no. 51 (December 2010): 22169–71, https://doi.org/10.1073/pnas.1011841108.

261 **Botanists are in a race to identify:** A. Antonelli, C. Fry, R. J. Smith, M. S. J. Simmonds, P. J. Kersey, H. W. Pritchard, M. S. Abbo, C. Acedo, J. Adams, A. M. Ainsworth, B. Allkin, W. Annecke, S. P. Bachman, K. Bacon, S. Bárrios, C. Barstow, A. Battison, E. Bell, K. Bensusan, M. I. Bidartondo, R. J. Blackhall-Miles, B. Bonglim, J. S. Borrell, F. Q. Brearley, E. Breman, R. F. A. Brewer, J. Brodie, R. Cámara-Leret, R. Campostrini Forzza, P. Cannon, M. Carine, J. Carretero, T. R. Cavagnaro, M.-E. Cazar, T. Chapman, M. Cheek, C. Clubbe, C. Cockel, J. Collemare, A. Cooper, A. I. Copeland, M. Corcoran, C. Couch, C. Cowell, P. Crous, M. da Silva, G. Dalle, D. Das, J. C. David, L. Davies, N. Davies, M. N. De Canha, E. J. de Lirio, S. Demissew, M. Diazgranados, J. Dickie, T. Dines, B. Douglas, G. Dröge, M. E. Dulloo, R. Fang, A. Farlow, K. Farrar, M. F. Fay, J. Felix, F. Forest, L. L. Forrest, T. Fulcher, Y. Gafforov, L. M. Gardiner, G. Gâteblé, E. Gaya, B. Geslin, S. C. Gonçalves, C. J. N. Gore, R. Govaerts, B. Gowda, O. M. Grace, A. Grall, D. Haelewaters, J. M. Halley, M. A. Hamilton, A. Hazra, T. Heller, P. M. Hollingsworth, N. Holstein, M.-J. R. Howes, M. Hughes, D. Hunter, N. Hutchinson, K. Hyde, J. Iganci, M. Jones, L. J. Kelly, P. Kirk, H. Koch, I. Krisai-Greilhuber, N. Lall, M. K. Langat, D. J. Leaman, T. C. Leão, M. A. Lee, I. J. Leitch, C. Leon, E. Lettice, G. P. Lewis, L. Li, H. Lindon, J. S. Liu, U. Liu, T. Llewellyn, B. Looney, J. C. Lovett, Ł. Łuczaj, E. Lulekal, S. Maggassouba, V. Malécot, C. Martin, O. R. Masera, E. Mattana, N. Maxted, C. Mba, K. J. McGinn, C. Metheringham, S. Miles, J. Miller, W. Milliken, J. Moat, J. G. P. Moore, M. P. Morim, G. M. Mueller, H. Muminjanov, R. Negrão, E. N. Lughadha, N. Nicolson, T. Niskanen, R. N. Womdim, A. Noorani, M. Obreza, K. O'Donnell, R. O'Hanlon, J.-M. Onana, I. Ondo, S. Padulosi, A. Paton, T. Pearce, O. A. P. Escobar, A. Pieroni, S. Pironon, T. A. K. Prescott, Y. D. Qi, H. Qin, C. L. Quave, L. Rajaovelona, H. Razanajatovo, P. B. Reich, E. Rianawati, T. C. G. Rich, S. L. Richards, M. C. Rivers, A. Ross, F. Rumsey, M. Ryan, P. Ryan, S. Sagala, M. D. Sanchez, S. Sharrock, K. K. Shrestha, J. Sim, A. Sirakaya, H. Sjöman, E. C. Smidt, D. Smith, P. Smith, S. R. Smith, A. Sofo, N. Spence, A. Stanworth, K. Stara, P. C. Stevenson, P. Stroh, L. M. Suz, E. C. Tatsis, L. Taylor, B. Thiers, I. Thormann, C. Trivedi, D. Twilley, A. D. Twyford, T. Ulian, T. Utteridge, V. Vaglica, C. Vásquez-Londoño, J. Victor, J. Viruel, B. E. Walker, K. Walker, A. Walsh, M. Way, J. Wilbraham, P. Wilkin, T. Wilkinson, C. Williams, D. Winterton, K. M. Wong, N. Woodfield-Pascoe, J. Woodman, L. Wyatt, R. Wynberg, and B. G. Zhang, *State of the World's Plants and Fungi 2020* (London: Royal Botanic Gardens, Kew, 2020), https://doi.org/10.34885/172.

261 **Almost 450 years of collecting:** Barbara M. Thiers, *The World's Herbaria 2019: A Summary Report Based on Data from Index Herbariorum,* New York Botanic Garden, January 10, 2020, http://sweetgum.nybg.org/science/docs/The_Worlds_Herbaria_2019.pdf.

266 **Nobel laureate James Watson wrote of Rosalind Franklin:** James D. Watson, *The Double Helix: A Personal Account of the Discovery of the Structure of DNA* (New York: Atheneum, 1968).

268 **Although women now represent:** "Doctoral Degrees Earned by Women, by Major," American Physical Society, accessed April 3, 2021, www.aps.org/programs/education/statistics/fraction-phd.cfm.

268 **A global survey of women:** Isabel Torres, Ryan Watkins, Martta Liukkonen, and Mei Lin Neo, "COVID Has Laid Bare the Inequities That Face Mothers in STEM," *Scientific American,* December 23, 2020, www.scientificamerican.com/article/covid-has-laid-bare-the-inequities-that-face-mothers-in-stem/.

269 **an economic study of gender differences:** Catherine Buffington, Benjamin Cerf, Christina Jones, and Bruce A. Weinberg, "STEM Training and Early Career Outcomes of Female and Male Graduate Students: Evidence from UMETRICS Data Linked to the 2010 Census," *American Economic Review* 106, no. 5 (May 2016): 333–38, https://doi.org/10.1257/aer.p20161124.

269 **42 percent of mothers and 15 percent of fathers leave:** Erin A. Cech and Mary Blair-Loy, "The Changing Career Trajectories of New Parents in STEM," *Proceedings of the National Academy of Sciences* 116, no. 10 (March 2019): 4182–87, https://doi.org/10.1073/pnas.1810862116.

271 **inclusion of codes of conduct:** Society for Economic Botany, accessed April 3, 2021, www.econbot.org/home/governance/code-of-conduct.html.

273 **I'd been photographed with my leg:** Ferris Jabr, "Could Ancient Remedies Hold the Answer to the Looming Antibiotics Crisis?," *New York Times Magazine,* September 18, 2016, www.nytimes.com/2016/09/18/magazine/could-ancient-remedies-hold-the-answer-to-the-looming-antibiotics-crisis.html.

Chapter 11. The One-Legged Hunter

297 **wild harvest enough food plants:** Cassandra L. Quave and Andrea Pieroni, "A Reservoir of Ethnobotanical Knowledge Informs Resilient Food Security and Health Strategies in the Balkans," *Nature Plants* 1 (February 2015): 14021, https://doi.org/10.1038/nplants.2014.21.

297 **The fermentation of wild fruits:** Andrea Pieroni, Renata Sõukand, Cassandra L. Quave, Avni Hajdari, and Behxhet Mustafa, "Traditional Food Uses of Wild Plants among the Gorani of South Kosovo," *Appetite* 108, no. 1 (January 2017): 83–92, https://doi.org/10.1016/j.appet.2016.09.024.

297 **coined the term *ethnozymology*:** Cassandra L. Quave and Andrea Pieroni, "Fermented Foods for Food Sovereignty and Food Security in the Balkans: A Case Study of the Gorani People of Northeastern Albania," *Journal of Ethnobiology* 34, no. 1 (March 2014): 28–43.

301 **declines in migratory birds:** Jonathan Franzen, "Last Song for Migrating Birds," *National Geographic,* July 2013, www.nationalgeographic.com/magazine/2013/07/songbird-migration/.

301 **there is little enforcement:** Daniel Ruppert, "Assessing the Effectiveness of the Hunting Ban in Albania," master's thesis, Eberswalde University of Sustainable Development, 2018.

302 **chewy ice cream to explosives:** Susanne Masters, Tinde van Andel, Hugo J. de Boer, Reinout Heijungs, and Barbara Gravendeel, "Patent Analysis as a Novel Method for Exploring Commercial Interest in Wild Harvested Species," *Biological Conservation* 243 (March 2020): 108454, https://doi.org/10.1016/j.biocon.2020. 108454.

302 **wild orchid tubers:** Amy Hinsley, Hugo J. de Boer, Michael F. Fay, Stephan W. Gale, Lauren M. Gardiner, Rajasinghe S. Gunasekara, Pankaj Kumar, Susanne Masters, Destario Metusala, David L. Roberts, Sarina Veldman, Shan Wong, and Jacob Phelps, "A Review of the Trade in Orchids and Its Implications for Conservation," *Botanical Journal of the Linnean Society* 186, no. 4 (April 2018): 435–55, https://doi.org/10.1093/botlinnean/box083.

304 **nineteen million gallons of defoliant chemicals:** Jackson Maogoto, "Inquiry on Complicity in War Crimes as Defined in Article 8(2) of the International Criminal Court," International Monsanto Tribunal in The Hague (October 2016): 5, https://www.monsanto-tribunal.org/upload/asset_cache/971889519.pdf?rnd= KxkE6d.

304 **Agent Orange defoliant destroyed forests:** Charles Ornstein, Hannah Fresques, and Mike Hixenbaugh, "The Children of Agent Orange," *ProPublica*, December 16, 2016, www.propublica.org/article/the-children-of-agent-orange.

307 **Lego Mindstorms robotic controller:** Marco Caputo, James T. Lyles, Monique S. Salazar, and Cassandra L. Quave, "LEGO MINDSTORMS Fraction Collector: A Low-Cost Tool for a Preparative High-Performance Liquid Chromatography System," *Analytical Chemistry* 92, no. 2 (January 2020): 1687–90, https://doi.org/ 10.1021/acs.analchem.9b04299.

308 **insect-repellent effects:** Charles L. Cantrell, Jerome A. Klun, Charles T. Bryson, Mozaina Kobaisy, and Stephen O. Duke, "Isolation and Identification of Mosquito Bite Deterrent Terpenoids from Leaves of American (*Callicarpa americana*) and Japanese (*Callicarpa japonica*) Beautyberry," *Journal of Agricultural and Food Chemistry* 53, no. 15 (July 2005): 5948–53, https://doi.org/10.1021/jf0509308.

308 **first to successfully isolate a crystal of high-enough quality:** Gina Porras, John Bacsa, Huaqiao Tang, and Cassandra L. Quave, "Characterization and Structural Analysis of Genkwanin, a Natural Product from *Callicarpa americana*," *Crystals* 9, no. 10 (September 2019): 491.

308 **clerodane diterpene, from the beautyberry:** Micah Dettweiler, Roberta J. Melander, Gina Porras, Caitlin Risener, Lewis Marquez, Tharanga Samarakoon, Christian Melander, and Cassandra L. Quave, "A Clerodane Diterpene from *Callicarpa americana* Resensitizes Methicillin-Resistant *Staphylococcus aureus* to β-Lactam Antibiotics," *ACS Infectious Disease* 6, no. 7 (June 2020): 1667–73, https://doi.org/ 10.1021/acsinfecdis.0c00307.

309 **growth of *Cutibacterium acnes*:** Rozenn M. Pineau, Sarah E. Hanson, James T. Lyles, and Cassandra L. Quave, "Growth Inhibitory Activity of *Callicarpa americana* Leaf Extracts against *Cutibacterium acnes*," *Frontiers in Pharmacology* 10 (October 2019): 1206, https://doi.org/10.3389/fphar.2019.01206.

309 **successful isolation of three active compounds:** Huaqiao Tang, Gina Porras, Morgan M. Brown, François Chassagne, James T. Lyles, John Bacsa,

Alexander R. Horswill, and Cassandra L. Quave, "Triterpenoid Acids Isolated from *Schinus terebinthifolia* Fruits Reduce *Staphylococcus aureus* Virulence and Abate Dermonecrosis," *Scientific Reports* 10 (May 2020): 8046, https://doi.org/10.1038/s41598-020-65080-3.

310 ***Kalanchoe fedtschenkoi* exhibited the most:** Nicholas Richwagen, James T. Lyles, Brandon L. F. Dale, and Cassandra L. Quave, "Antibacterial Activity of *Kalanchoe mortagei* and *K. fedtschenkoi* against ESKAPE Pathogens," *Frontiers in Pharmacology* 10 (February 2019): 67, https://doi.org/10.3389/fphar.2019.00067.

Chapter 12. Cassandra's Curse

312 **total drug-resistant TB:** H. M. Adnan Hameed, Md. Mahmudul Islam, Chiranjibi Chhotaray, Changwei Wang, Yang Liu, Yaoju Tan, Xinjie Li, Shouyong Tan, Vincent Delorme, Wing W. Yew, Jianxiong Liu, and Tianyu Zhang, "Molecular Targets Related Drug Resistance Mechanisms in MDR-, XDR-, and TDR-*Mycobacterium tuberculosis* Strains," *Frontiers in Cellular and Infection Microbiology* 8 (April 2018): 114, https://doi.org/10.3389/fcimb.2018.00114.

313 **550,000 cases of gonorrhea:** Centers for Disease Control and Prevention, *Sexually Transmitted Disease Surveillance 2018*, October 1, 2019, https://doi.org/10.15620/cdc.79370.

313 **While the drug-discovery pipeline:** Stephanie N. Taylor, David H. Morris, Ann K. Avery, Kimberly A. Workowski, Byron E. Batteiger, Courtney A. Tiffany, Caroline R. Perry, Aparna Raychaudhuri, Nicole E. Scangarella-Oman, Mohammad Hossain, and Etienne F. Dumont, "Gepotidacin for the Treatment of Uncomplicated Urogenital Gonorrhea: A Phase 2, Randomized, Dose-Ranging, Single-Oral Dose Evaluation," *Clinical Infectious Diseases* 67, no. 4 (August 2018): 504–12, https://doi.org/10.1093/cid/ciy145; John O'Donnell, Ken Lawrence, Karthick Vishwanathan, Vinayak Hosagrahara, and John P. Mueller, "Single-Dose Pharmacokinetics, Excretion, and Metabolism of Zoliflodacin, a Novel Spiropyrimidinetrione Antibiotic, in Healthy Volunteers," *Antimicrobial Agents and Chemotherapy* 63 (October 2018): e01808-01818, https://www.ncbi.nlm.nih.gov/pmc/articles/PMC6325203/.

317 **review papers on plant-derived antibiotics:** Gina Porras, François Chassagne, James T. Lyles, Lewis Marquez, Micah Dettweiler, Akram M. Salam, Tharanga Samarakoon, Sarah Shabih, Darya Raschid Farrokhi, and Cassandra L. Quave, "Ethnobotany and the Role of Plant Natural Products in Antibiotic Drug Discovery," *Chemical Reviews* 121, no. 6 (November 2020): 3495–3560, https://doi.org/10.1021/acs.chemrev.0c00922; François Chassagne, Tharanga Samarakoon, Gina Porras, James T. Lyles, Micah Dettweiler, Lewis Marquez, Akram M. Salam, Sarah Shabih, Darya Raschid Farrokhi, and Cassandra L. Quave, "A Systematic Review of Plants with Antibacterial Activities: A Taxonomic and Phylogenetic Perspective," *Frontiers in Pharmacology* 11 (January 2020): 586548, https://doi.org/10.3389/fphar.2020.586548.

321 **one and two people per every one hundred thousand:** "Guillain-Barré Syndrome," Rare Disease Database, National Organization for Rare Disorders, accessed April 4, 2021, https://rarediseases.org/rare-diseases/guillain-barre-syndrome/.

322 **An alarming 30 percent of Guillain-Barré patients:** Mary-Anne Melone, Nicholas Heming, Paris Meng, Dominique Mompoint, Jérôme Aboab, Bernard Clair, Jérôme Salomon, Tarek Sharshar, David Orlikowski, Sylvie Chevret, and Djillali Annane, "Early Mechanical Ventilation in Patients with Guillain-Barré

Syndrome at High Risk of Respiratory Failure: A Randomized Trial," *Annals of Intensive Care* 10 (September 2020): 128, https://doi.org/10.1186/s13613-020-00742-z.

322 **9–13 percent of these patients die:** Atul Ashok Kalanuria, Wendy Zai, and Marek Mirski, "Ventilator-Associated Pneumonia in the ICU," *Critical Care* 18 (March 2014): 208, https://doi.org/10.1186/cc13775.

322 **My knowledge of the antibiotic-resistant:** Su Young Chi, Tae Ok Kim, Chan Woo Park, Jin Yeong Yu, Boram Lee, Ho Sung Lee, Yu Il Kim, Sung Chul Lim, and Yong Soo Kwon, "Bacterial Pathogens of Ventilator-Associated Pneumonia in a Tertiary Referral Hospital," *Tuberculosis and Respiratory Diseases* 73, no. 1 (2012): 32–37, https://doi.org/10.4046/trd.2012.73.1.32.

334 **physicians who specialize in infectious diseases:** Timothy Sullivan, "What Is the 'Relative Value' of an Infectious Disease Physician?," *Health Affairs*, February 3, 2017, https://www.healthaffairs.org/do/10.1377/hblog20170203.058600/full/.

334 **Some experts are calling for a subscription model:** "Do We Need a Netflix for Antibiotics?," *Financial Times*, video, 4:50, February 9, 2021, www.ft.com/video/adada10f-5747-4976-a3e0-958b0165e0ef.

335 **"valley of death":** Attila A. Seyhan, "Lost in Translation: The Valley of Death across Preclinical and Clinical Divide—Identification of Problems and Overcoming Obstacles," *Translational Medicine Communications* 4 (November 2019): 18, https://doi.org/10.1186/s41231-019-0050-7.

Epilogue

340 **born with spina bifida:** "Spina Bifida and Agent Orange," US Department of Veterans Affairs, accessed April 2, 2021, www.publichealth.va.gov/exposures/agentorange/birth-defects/spina-bifida.asp.